star*

◆ science, technology and reading – a resource for teachers

acknowledgements

Acknowledgements

Viv Aylwood, PGCE student, University of Sussex
Nathan Barlex
Peter Borrows, CLEAPSS
Wenda Bradley, Headteacher, Somerhill Junior School, Hove
David Brown, Headteacher and staff, Lodge Mount Primary School, Loughborough
Caroline Evans, Educational Communications Ltd
Mary Hinton, Somerhill Junior School, Hove
Martin Hollins
Lesley McGorrigan, Somerhill Junior School, Hove
Jan Rees, Royal Society of Chemistry Teacher Fellow 1998-1999
Franca Reid, Headteacher and staff, Longforgan Primary School, Scotland
John Reynolds, ESTA
Lisa Smith, Wood End Junior School, Greenford
Melinda Stone, Westdene Primary School, Brighton
Students from the Reading College of Art and Design
Dr Stephen Turner, Royal Society of Chemistry
The staff, Bradley Church of England School, Huddersfield
The staff, Christ Church Woodhouse, Church of England School, Huddersfield
The staff, Fixby Junior and Infant School, Huddersfield
Rose Watkin, Deputy Headteacher and staff, Sidlesham County Primary School, Chichester
David Williams, University of Sussex
Mike Willson, the PGCE science team and students, University of Sussex

Published by the Association for Science Education,
College Lane, Hatfield, Herts AL10 9AA

Website: http://www.ase.org.co.uk

Tel: 01707 283000
Fax: 01707 266532

Edited by **Evelyn van Dyk**
Layout by **Commercial Campaigns**

Printed by Streets Printers, Royston Road, Baldock, SG7 6NW

ISBN 0 86357 315 0

Contents

The star* story

How many children have ever been invited to sing and dance to science? To view it lyrically? To breathe poetic life into it?

A star* is born

A sparkling combination of science and poetry is **star*** – Science, Technology and Reading – with a wildcard asterisk to signify its unlimited potential in bringing the arts and sciences together, and fostering imaginative cross-curricular teaching.

In a close collaboration of like-minded organisations – The Royal Society of Chemistry, Esso, and the Design Council together commissioned Michael Rosen, foremost childrens' poet in the UK, to write a hundred poems covering scientific concepts in chemistry, physics, the environment, and design and technology, with the latter drawing on the Design Council's Millennium Products – a showcase of British innovative design. The Institute of Physics was also involved in the project.

This book of poems, complete with background scientific knowledge and detailed suggestions for suitable educational activities, is being published as a handbook by the Association of Science Education. It includes colourful illustrations by students of Reading College of Art and Design. The poems will also be available with a commentary by the poet in a separate complete edition called *Centrally Heated Knickers*, to be published by Puffin in Autumn 2000.

Is there any reason that the arts and sciences cannot feed into and inspire each other? The handbook is not intended to be prescriptive, but rather to encourage new ways of teaching across the curriculum. It is hoped that teachers using it will come up with their own innovations allowing students to view subjects from different perspectives, thereby enhancing their creative and thinking skills and bolstering their enthusiasm to learn.

The star* group

The Royal Society of Chemistry is the professional body for chemists and the learned society for chemistry with 46,000 members world-wide in industry and academia. The Society confronts issues affecting the future of chemists and chemistry and is a driving force behind chemistry education at all levels. It also promotes chemistry to the public and takes a leading role in chemistry throughout Europe. It is proactive in supporting the curriculum and so when the National Literacy Strategy was introduced, the Society was inspired, along with the other **star*** partners, to investigate the synergistic effect of teaching science through literature and particularly poetry.

Esso is a leading integrated oil company operating in the UK. It has long supported education at all levels with a focus on subjects relevant to its business, particularly science, technology, maths and environmental education. It seeks to encourage better teaching in order to inspire students to become and remain enthusiastic enough to choose careers in these areas. Poetry offers a novel way of fulfilling this aim, and is particularly relevant given the Government focus on literacy in primary schools. The **star*** project meets all these objectives in an innovative way.

The Design Council aims to inspire the best use of design for the benefit of society and the economy in the UK. Part of its remit is to encourage innovative approaches to education. The **star*** project provides just such a fresh way of teaching subjects which feed into design and technological creativity, being itself an example of innovative thinking. The 25 poems relating to design and technology have drawn on the Design Council's Millennium Products initiative, a show-case of UK design creativity in all sectors.

The Institute of Physics is an international learned and professional society for the advancement and dissemination of physics, pure and applied, and the promotion of physics education through supporting and presenting physics research and understanding to other scientists, decision-makers and the public. The Institute has over 26,000 members world-wide. It has supported the **star*** project in the belief that physics needs to be seen as an activity that humans take part in. Using poetry is one way to do this, allowing students to explore scientific concepts in another context, thereby helping relate physics to their wider experience of the world around them.

The aim of this collection is to illustrate how science and poetry make excellent partners and that through poetry children can access a range of scientific ideas and ways of working.

This collection of science poems and activities is not a scheme but designed to be used by teachers in a variety of ways, for example:

- as a starting point for science thinking and activity;
- to encourage children to write their own science based poems;
- to make links between science and literacy;
- to teach and develop a variety of literacy skills, from past tense to alliteration.

Each unit comprises of the following:

Poem as text

Here the poem is plain text which could be photocopied and used with children on OHP transparency or individual sheets.

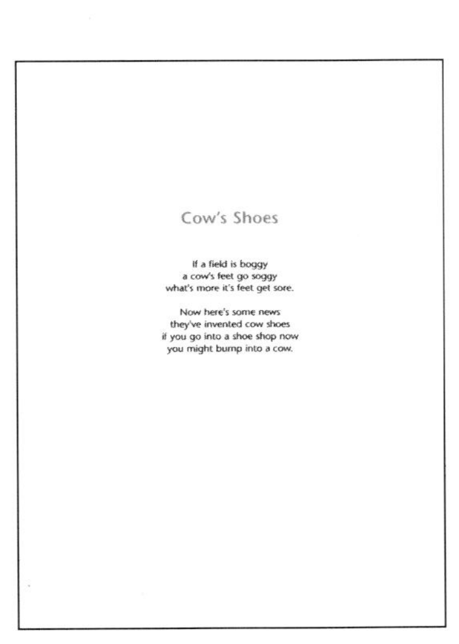

Cow's Shoes

If a field is boggy
a cow's feet go soggy
what's more it's feet get sore.

Now here's some news
they've invented cow shoes
if you go into a shoe shop now
you might bump into a cow.

Colour illustration

On this page the poem has been illustrated and could be copied and laminated or copied and used as an OHP transparency.

Teacher notes

This section begins with a list of questions and discussion points that can be used to start a debate. It includes the science background related to the content of the poem and follow up work in science-related activities. Science and key skills are also identified to assist the teacher.

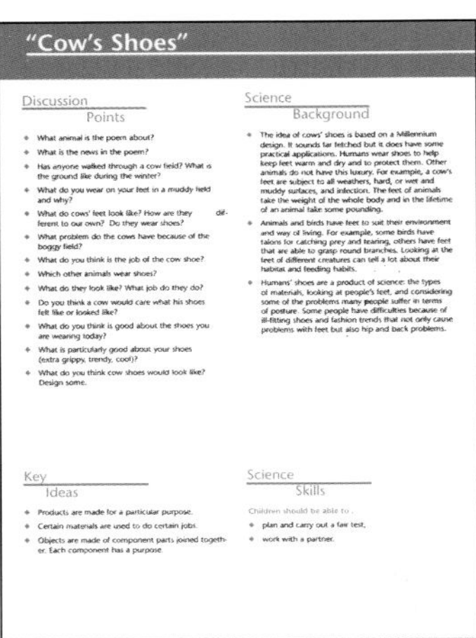

"Cow's Shoes"

Discussion Points

- What animal is the poem about?
- What is the news in the poem?
- Has anyone walked through a cow field? What is the ground like during the winter?
- What do you wear on your feet in a muddy field and why?
- What do cows' feet look like? How are they different to our own? Do they wear shoes?
- What problem do the cows have because of the boggy field?
- What do you think is the job of the cow shoe?
- Which other animals wear shoes?
- What do they look like? What job do they do?
- Do you think a cow would care what his shoes felt like or looked like?
- What do you think is good about the shoes you are wearing today?
- What is particularly good about your shoes (extra grippy, trendy, cool)?
- What do you think cow shoes would look like? Design some.

Science Background

- The idea of cows' shoes is based on a Millennium design. It sounds far fetched but it does have some practical applications. Humans wear shoes to help keep feet warm and dry and to protect them. Other animals do not have this luxury. For example, a cow's feet are subject to all weathers, hard, or wet and muddy surfaces, and infection. The feet of animals take the weight of the whole body and in the lifetime of an animal take some pounding.
- Animals and birds have feet to suit their environment and way of living. For example, some birds have talons for catching prey and tearing, others have feet that are able to grasp round branches. Looking at the feet of different creatures can tell a lot about their habitat and feeding habits.
- Humans' shoes are a product of science: the types of materials, looking at people's feet, and considering some of the problems many people suffer in terms of posture. Some people have difficulties because of ill-fitting shoes and fashion trends that not only cause problems with feet but also hip and back problems.

Key Ideas

- Products are made for a particular purpose.
- Certain materials are used to do certain jobs.
- Objects are made of component parts joined together. Each component has a purpose.

Science Skills

Children should be able to :

- plan and carry out a fair test,
- work with a partner.

42 ◆ "Cow's Shoes" ◆ star* designs ◆

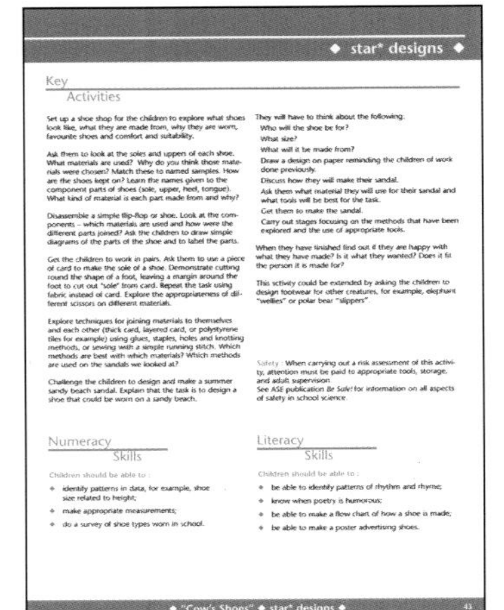

◆ star* designs ◆

Key Activities

Set up a shoe shop for the children to explore what shoes look like, what they are made from, why they are worn, favourite shoes and comfort and suitability.

Ask them to look at the soles and uppers of each shoe. What materials are used? Why do you think those materials were chosen? Match these to named samples. How are the shoes kept on? Learn the names given to the component parts of shoes (sole, upper, heel, tongue). What kind of material is each part made from and why?

Disassemble a simple flip-flop or shoe. Look at the components – which materials are used and how were the different parts joined? Ask the children to draw simple diagrams of the parts of the shoe and to label the parts.

Get the children to work in pairs. Ask them to use a piece of card to make the sole of a shoe. Demonstrate cutting round the shape of a foot, leaving a margin around the foot to cut out "sole" from card. Repeat the task using fabric instead of card. Explore the appropriateness of different scissors on different materials.

Explore techniques for joining materials to themselves and each other (thick card, layered card, or polystyrene tiles for example) using glues, staples, holes and knotting methods, or sewing with a simple running stitch. Which methods are best with which materials? Which methods are used on the sandals we looked at?

Challenge the children to design and make a summer sandy beach sandal. Explain that the task is to design a shoe that could be worn on a sandy beach.

They will have to think about the following:

Who will the shoe be for?

What size?

What will it be made from?

Draw a design on paper reminding the children of work done previously.

Discuss how they will make their sandal.

Ask them what material they will use for their sandal and what tools will be best for the task.

Get them to make the sandal.

Carry out stages focusing on the methods that have been explored and the use of appropriate tools.

When they have finished find out if they are happy with what they have made? Is it what they wanted? Does it fit the person it is made for?

This sctivity could be extended by asking the children to design footwear for other creatures, for example, elephant "wellies" or polar bear "slippers".

Safety : When carrying out a risk assessment of this activity, attention must be paid to appropriate tools, storage, and adult supervision.
See ASE publication *Be Safe!* for information on all aspects of safety in school science.

Numeracy Skills

Children should be able to :

- identify patterns in data, for example, shoe size related to height;
- make appropriate measurements;
- do a survey of shoe types worn in school.

Literacy Skills

Children should be able to :

- be able to identify patterns of rhythm and rhyme;
- know when poetry is humorous;
- be able to make a flow chart of how a shoe is made;
- be able to make a poster advertising shoes.

◆ "Cow's Shoes" ◆ star* designs ◆ 43

Teachers notes

The activities are suggestions for work related to the poem in science. The numeracy and literacy statements suggest links between the poem and these areas.

Acorn, Conker and Key

Hey acorn!
Who d'you think you are?
A hard guy?
You look like a hard-boiled egg
sitting in a cup
Well I'm telling you,
hard guy,
a squirrel is going to find you,
and if he doesn't eat you
he's going to bury you.

Just because you're the start-out point
for the tree
that made the first ships
to go right round the world
doesn't mean you're a bit shot.
You're no circum-navigator!

And you, Conker!
Always up for a fight, aren't you?
"Let me at 'im
Let me at 'im"-
that's you!

Well, I know your little secret:
there you are,
lying about in your little green house
and then, when the walls split,
out you pop
like you think
you're some shiny new car
cruising out of the garage.

But I've seen inside
your little green house:
you lie for weeks
all tucked up in a soft white bed,
don't you?
Hard man! Huh!

But you sycamore key,
you're plain weird, plain weird.
You jive about in the air,
jiving in the wind;
cool moves, man!

But then,
you lie on the ground
in a heap
with all the other jivers
looking like a dead moth.
There's hundreds of
you dead moths
lying there.
You're weird.
Where are you at?

Related poem:

Growing Apples

Acorn, Conker and Key

Hey Acorn!
Who d'you think you are?
A hard guy?
'ou look like a hard-boiled egg
tting in a cup.
ell I'm telling you,
d guy,
uirrel is going to find you,
if he doesn't eat you
going to bury you.

because you're the start-out point
ne tree
made the first ships
o right roung the world
esn't mean you're a big shot.
'ou're no circum - navigator!

And you, Conker!
Always up for a fight, aren't you?
'Let me at 'im
Let me at 'im'-
that's you!

Well, I know your little secret:
there you are,
lying about in your little green house
and then, when the walls split,
out you pop
like you think
you're some shiny new car
cruising out of the garage.

But I've seen inside
your little green house:
you lie for weeks
all tucked up in a soft white bed,
don't you?
Hard man! Huh!

But you sycamore key,
you're plain wierd, plain wierd.
You jive about in the air,
jiving in the wind;
cool moves, man!

But then,
you lie on the ground
in a heap
with all the other jivers
looking like a dead moth.
There's hundreds of
you dead moths
lying there.
You're wierd.
Where are you at?

"Acorn, Conker and Key"

Discussion Points

It would be helpful to have an acorn, a conker and a sycamore key to show to the children, preferably still attached to a piece of twig, or at least good photographs of them.

- What is an acorn? Does it look like "a little hard-boiled egg sitting in a cup"?
- What do you know about squirrels and acorns? Is the poem right?
- Why do you think the poem says the acorn is "the start-out point for the tree that made the first ships to go right round the world"? What is a circumnavigator?
- What is a conker? Why is the conker "always up for a fight"?
- What is the conker's little green house? How do the walls split? Why is the conker like some shiny new car? What is the conker's soft white bed?
- What is a sycamore key? Why do you think it's weird? Does it jive about in the air? Why?
- Have you seen sycamore keys lying on the ground? Did they look like dead moths?
- What do you think the sycamore key might be "at", lying on the ground?
- What will happen if you plant an acorn, sycamore key and conker? Describe the life cycle of one of them.

Science Background

- Most people are familiar with the process of plant growth and the plant life cycle, but there is a widespread misconception – particularly among children – about how plants obtain their food. Plants make their own food from water, carbon dioxide from the air and sunlight. This process is called photosynthesis and takes place mainly in their leaves using a green substance called chlorophyll.
- Seeds contain a supply of food, sufficient for the plant to germinate and grow its first leaves which start to photosynthesise, so the plant becomes self-sufficient.
- Despite what the gardening programmes and adverts say, plants do not get food directly from the soil or fertilizers. Plants do need water, which they take in mainly through their roots, and small amounts of other nutrients, such as nitrogen and sulphur (in soluble form), which are also absorbed through the roots and used in other processes within the plant such as cell formation. The role of fertilizers is to enrich the nutrients in the soil.
- Flowering plants reproduce sexually, the flower containing both the male and female parts. For fertilisation to happen, pollen from the male part has to be transferred to the female part, usually of another flower on another plant of the same type. The commonest ways for this to take place are by insects or the wind. After fertilisation has occurred, the female part develops into the fruit which contains the seed(s). The seed contains the embryo of a new plant and a supply of food, so that it can germinate when it finds itself in the right conditions. Children do not always realise that trees have flowers which produce seeds and fruit.

Key Ideas

- Plants grow from seeds, which come in various shapes and sizes.
- Plants show seasonal changes.
- The life cycle of a flowering plant.

Science Skills

Children should be able to :

- understand and write instructions;
- measure and compare accurately;
- consult books and data-bases;
- organise information and findings for report and display.

Key Activities

With the children, look at a small selection of seeds of different shapes and sizes such as cress, peas, mung beans, broad beans, sunflower seeds, acorns, conkers and apple seeds. Ask the children what they think they are, where they came from and what we might do with them.

Then, in groups, plant a sample of each sort of seed to germinate. Place compost in a transparent container, for example, a cut down plastic lemonade bottle. Plant the seed in the compost against the wall of the container so that children can watch it germinate. Observe the seeds once a day for at least a week. Make sure they don't dry out. Tell the children to make drawings of what happens.

Ask the children to explain what is happening, and to suggest what might happen if the seeds were left for a longer time – they could draw predictions of the next four weeks. Older children could cultivate a small school garden, growing vegetables and flowers, and keep a log and take photographs, perhaps using a digital camera.

Children could observe and record the changes to the plants around the school in the spring, summer, autumn and winter, and predict what was going to happen next. As well as writing and drawing, they could make seasonal displays of, say, leaves, flowers, fruits or seeds.

Older children could research the plants and the changes, particularly the ways in which fertilisation occurred in the flowers.

Small groups of children could be responsible for a particular plant, perhaps taking it home during holidays and/or observing similar plants at home.

The same approach would be appropriate but with closer attention to the detail of important stages such as the production of the leaf, flower and fruit. This should be supported by resources such as books or data-bases, visits to gardens or nurseries and talks by invited speakers – perhaps including an experienced parent-gardener.

Children could focus particularly on the fertilisation process and how the seeds are dispersed, making appropriate use of secondary resources. Challenge children to tell someone else the "story" of a life cycle using an interesting way of communication, such as role play.

For more information on growing trees from seed, please contact "Growing with Trees" project on 01962 846 258 or www.totap.org.uk website.

Safety : Children must take care when working close to and handling all living things, including plants. They must wash their hands after contact; parts of some plants are irritants and may also be poisonous. You need to be alert to allergies or allergic reactions. See ASE publication *Be Safe!* for information on all aspects of safety in school science.

Numeracy Skills

Children should be able to :

- measure small objects of differing sizes (seeds);
- use data in different ways;
- make comparisons;
- calculate mean, mode and median.

Literacy Skills

Children should be able to :

- describe from observations;
- use appropriate vocabulary;
- make notes about seed growth;
- make flow charts explaining seed growth.

Berlam Bam Boola

That night
we made a fire
on the beach
and Alan danced about
singing:
Berlam bam boola
Berlam bam boola
tooty fruity

I found
a bit of drift wood
that looked like
a cow's skull
and it burned up bright
Berlam bam boola
Berlam bam boola
tooty fruity

When we found
the remains of the fire,
in the ashes
I could just make out
the shape of a cow's skull.

Berlam Bam Boola

That night
we made a fire
on the beach
and Alan danced about
singing:
Berlam bam boola
Berlam bam boola
tooty fruity

I found
a bit of drift wood
that looked like
a cow's skull
and it burned up bright
Berlam bam boola
berlam bam boola
tooty fruity

When we found
the remains of the fire,
in the ashes
I could just make out
the shape of a cow's skull.

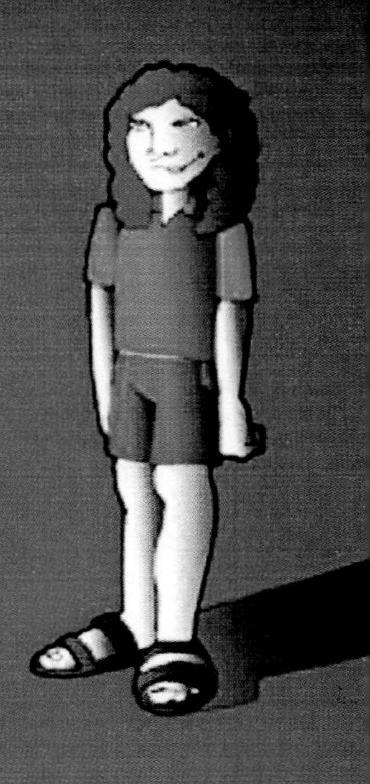

Discussion Points

- What did they burn on the fire?
- What is driftwood? Have you ever collected it? Where did you find it?
- In the poem, could they get the wood back after they burnt it? Could they reverse the change?
- What else do you know that is an irreversible reaction?
- What was left after the burning?
- Was a new material made?
- Do you think the beach fire was safe? How would you make it safe?

Science Background

- Burning requires three factors, fuel, oxygen and a source of heat. It is irreversible.
- In a chemical reaction a new substance is formed as a result, for example: ash, carbon dioxide, water. Also a flame is often seen. Burning a candle is a good example of heating and burning. Lighting the wick causes it to burn and the wax to melt; the wax vaporises then this burns. There have been both physical and chemical changes. If the candle is burned in an inverted jar, condensation is seen because water is produced as a result of burning. As oxygen is needed, the larger the jar or the fewer the candles, the longer the burning will continue. If the jar and candle are stood in water, as the candle burns, water rises inside the jar. This is for a variety of complex chemical and physical reasons including: the fact that oxygen is being used up; there is a difference in pressure between the inside and outside of the jar causing water to enter it, and the volume of a gas depends on temperature.

Key Ideas

- Changes that occur when most materials burn are not reversible.
- New materials are formed.
- Burning requires fuel, oxygen and a source of heat.

Science Skills

Children should be able to :

- follow instructions carefully;
- handle materials carefully;
- use a stopwatch accurately;
- suggest safety procedures;
- work co-operatively in pairs or groups.

Key Activities

Using a nightlight as a heat source, ask the children to burn a variety of materials, for example, paper, wooden spill or twig, wire wool, cotton, small pieces of fabric. Carefully observe what happens. Are any new materials formed? Metal tongs should be used to hold the burning materials or let them burn in small, metal dishes. If a candle is used, stand it in a tray of sand.

Heating a material causes a temperature rise and often results in a change of state. Burning is setting alight and a flame is usually seen; new materials are formed. Burning needs oxygen.

Carefully watch a candle burn. Observe the wax and the wick. What is happening? Now burn the candle and cover with a jar. Closely observe what happens. Time how long the flame stays alight. What happens if two candles are burned in the jar? Investigate how a candle burns in different-sized jars. After the investigation, ask the children to explain their results.

Why does the candle lasts longer in the bigger jar? Carry out a teacher demonstration to show that oxygen is used up. Stand the candle in a tray of water, with the jar over it. The candle burns, then goes out and water rises up inside the jar.

Discuss pollution of the atmosphere and the destruction of forests by burning.

Safety : Wear safety goggles during burning experiments. Stand the candles in sand or water. Do not touch hot materials. Some materials, for example, some plastics, give toxic fumes when burned. Use small samples, for example, 1 cm^2.
See ASE publication *Be Safe!* for information on all aspects of safety in school science.

Numeracy Skills

Children should be able to :

- use a stop watch;
- collect data in a table;
- represent data as a line graph;
- use numerical data to draw conclusions.

Literacy Skills

Children should be able to :

- create a headline relating to something burning or a fire;
- know the format of a newspaper article;
- write an article about a fire using a journalistic style.

Chippy Breath

After football
my dad buys me fish and chips
and my hot chippy breath
makes clouds in the air
and rain on the windows
of the bus
all the way home

I write the score
on the wet glass
- but only when we win.

Related poems:

Hot Pants
Thirsty Land

Chippy Breath

After football
my dad buys me fish and chips
and my hot chippy breath
makes clouds in the air
and rain on the windows
of the bus
all the way home.

I write the score
on the wet glass
- but only when we win.

"Chippy Breath"

Discussion Points

- Why does his or her breath make clouds in the air?
- What do we breathe out?
- Why does our breath sometimes make clouds in the air?
- What does that tell you about the air in the bus?
- What is the "rain" on the bus window?
- Why has his breath turned to water on the window?
- Where else have you seen this happen?
- There is a "water cycle" happening in the bus. How would you describe it to someone else?

Science Background

- Evaporation occurs when a liquid appears to "dry up".
- It does not need to be heated to a high degree for the liquid state to change to the gaseous state. Those molecules on the surface of the liquid, which have more energy, change to a colourless gas called water vapour, and move from the liquid into the air.
- The more liquid that is exposed to the air, the faster it will evaporate.
- When water is boiled, the gas phase is very hot (100ºC).
- The gas is called steam and it is invisible.
- When water vapour or steam hits the colder air, the gaseous state turns back to the liquid state (condenses) and forms tiny droplets in the air.
- These droplets remain suspended as a fog, mist or cloud. If these tiny droplets hit a cold surface they coalesce further to form water.
- Evaporation occurs at low temperatures but increases as the temperature rises because the surface molecules have more energy and move away faster. As evaporation occurs it has a cooling effect on the remaining liquid.
- A cloud is the name for a volume of air well above ground where water vapour has condensed to form tiny water droplets or ice crystals. The rising air prevents them from falling, but as they grow bigger a point is reached where they are too heavy to be held in the air and they fall as rain, sleet or snow. This is called precipitation.

Key Ideas

- That water changes from a liquid to a gas called water vapour, which is in the air (evaporation).
- Evaporation is affected by the area of the surface from which the liquid is evaporating, and also by temperature.
- Evaporation can be reversed when the water vapour cools. This is called condensation.
- Evaporation and condensation are part of the water-cycle.

Science Skills

Children should be able to :

- plan a fair test;
- following instructions correctly;
- use apparatus carefully;
- use a thermometer properly;
- work co-operatively in pairs or groups.

Key Activities

Leave containers of water around the classroom to evaporate and observe what happens. Investigate whether temperature affects evaporation.

Ask the children to find out whether different liquids evaporate at the same rate, for example, vinegar or lemon juice.

Leave identical containers in places of different temperatures and measure the rate of evaporation. Older children can plot a line graph relating to temperature and amount evaporated over time.

Investigate whether the size of the surface area affects rate of evaporation. Children can use containers with openings of a different surface area which they can measure by drawing round on squared paper. They must keep the amount of water and temperature constant. Again, older children can plot a line graph.

Get the children to breathe on cold mirrors and discuss where the water droplets come from.

Demonstrate condensation with a boiling kettle and mirror. If you insert a thermometer just above the spout into the invisible steam, the temperature shoots to 100ºC or above and drops as you move it into the visible water vapour.

Children are sometimes confused as to what steam is. As steam is an invisible gas, they need to be aware that the water droplets (condensation) they see when steam hits cooler air is water vapour and not the steam itself.

A water-cycle can be set up in a variety of ways. A beaker of water in a sealed polythene bag left in a warm place is the simplest. This can also be demonstrated by standing a dish of ice over a beaker of warm water. As the evaporated water vapour hits the bottom of the saucer, it collects and eventually falls in drops.

The part played by plant life can be demonstrated by putting a pot plant in a polythene bag. Eventually the polythene bag will cloud over due to condensation. This illustrates that plants give out water even though we cannot see it.

Safety : Be aware of the hazards when using thermometers and mirrors. The glass edges of mirrors must be covered. Take care when handling liquids. See ASE publication *Be Safe!* for information on all aspects of safety in school science.

Numeracy Skills

Children should be able to :

- collect data;
- measure area accurately;
- produce a table;
- create a line graph;
- interpret a line graph.

Literacy Skills

Children should be able to :

- write from their own experience;
- recognise how poets use words;
- read and interpret moods and feelings in poems.

Concrete Paw

Our friends next door
have moved away.
They put everything
in a van today.

They got in a car
and went.
That's it.
No one's left.

My friends gone.
Her sister's gone.
Their mum's gone.
Their dog Sniffy's gone.

Everyone's gone.
Nothing left...

... except for
the mark of Sniffy's paw
printed into the concrete
by their front door
where it'll stay
forever more.

Concrete Paw

Our friends next door
have moved away.
They put everything
in a van today.

They got in a car
and went.
That's it.
No one's left.

My friends gone.
Her sister's gone.
Their mum's gone.
Their dog Sniffy's gone.

Everyone's gone.
Nothing left...

...except for
the mark of Sniffy's paw
printed into the concrete
by their front door
where it'll stay
forever more.

"Concrete Paw"

Discussion Points

- What do you think happens when someone moves house?
- Have you seen anyone move, and use a removal van?
- What kind of things do people leave behind that they can't take with them?
- What was left behind in this poem?
- What is concrete? What it is made of? Do you know how it is made?
- How is a concrete path made?
- What is the consistency of concrete when you mix it?
- How does concrete change when you are making it?
- What other things can you make with it?
- How do you think Sniffy's paw mark got into the concrete?
- What can you do to concrete while it is drying?

Science Background

- A chemical change is one that results in the formation of a new material, and is generally very difficult to reverse. This can be seen by cooking and burning. Some chemical reactions, such as making concrete, are very useful; others like rusting can be a serious and expensive problem. At primary level the nature of chemical change is difficult to teach. The new material produced is not always obvious. At this level it is appropriate to keep within the experience of the children.
- Some examples of chemical changes can be shown to children, for example, cooking: boiling, frying an egg, baking cakes, biscuits or bread. Making cheese and yoghurt is a change brought about not by heat but by the production of acid by bacteria. Similarly vinegar with warm milk will cause it to change chemically and curdle, a reaction which produces casein, a plastic-like substance. Baking powder in hot water or sodium bicarbonate and vinegar also create chemical changes.
- Cement is made by heating crushed limestone and clay. It is mixed with sand and water for use in building. The compounds in the cement react with water and set hard. Plaster of Paris (modelling plaster) produces heat when it is mixed with water, sets hard and produces a new material.
- Many chemical reactions involve oxygen, some of which are familiar to children, such as burning and rusting. Rusting is a reaction in which iron combines with oxygen and water to give the red/brown iron oxide, the process is called oxidation.

Key Ideas

- Mixing things can cause them to change.
- Some changes cannot be reversed.
- Useful products can be made from chemical reactions.
- Some chemical reactions are not useful.

Science Skills

Children should be able to :

- plan a fair test;
- choose and use apparatus with care and use it properly;
- read a thermometer accurately;
- carry out instructions carefully;
- observe reactions and record results accurately;
- work co-operatively in pairs or groups.

Key Activities

Ask the children to follow instructions for making plaster of Paris models using a mould. This could fit in with an art or history topic.

Making a fossil:

1 Press a shell into plasticine to make a shape;

2 Using a paperclip and card strip, make a collar round the shape;

3 Make a cream with two tablespoons of plaster of Paris (modelling plaster) and water;

4 Pour into the mould and leave for 20 minutes, or until set;

5 Separate the fossil from the mould.

During this chemical reaction, there is a rise in temperature. Whilst making the model, get the children to set aside some mixed plaster of Paris in a small container. Wrap the end of a thermometer in foil. Take the temperature of the plaster as it sets.

Ask children whether they think making plaster of Paris is a reversible process. Some children could investigate this. Get them to suggest a method. Crush the made-up plaster and mix with water to see if it will set again. If sufficient water was used in the first setting, it will not set again as the first was a permanent, chemical change.

Some chemical changes are not useful, for example, rusting. Find out if the children know what causes rust. Investigate the conditions for rusting. See if the children can plan this for themselves. If not, suggest iron or steel nails to be put in screw top jars or polythene bags. Use dry nails, wet nails or nails covered in Vaseline to waterproof them.

Discuss the presence of air/oxygen. It is very difficult to eliminate this but it sometimes works if you boil the air out of the water, put the boiled water in the jar with the nail, quickly cover with oil and screw on the top. If you are using a polythene bag, suck the air out of the bag and tie the top.

Having discovered the conditions for rusting, the children might also investigate which materials rust. Test a variety of small objects of different materials, including other metals. Investigate whether all liquids cause rusting. Test a variety of liquids: oil, cola, lemonade, vinegar, water, lemon juice. These tests can be planned by the children.

Safety : Wear goggles for crushing the plaster. Beware asthmatics breathing dry plaster. It may also irritate sensitive skin. Crushing plaster can be done more safely by first putting it in a polythene bag. Use a rolling pin. Beware the points on nails and using glass jars and thermometers.
See ASE publication *Be Safe!* for information on all aspects of safety in school science.

Numeracy Skills

Children should be able to :

- read a thermometer;
- measure ingredients;
- use the scales on a thermometer or weighing scales, etc.

Literacy Skills

Children should be able to :

- read narrative poems;
- perform a poem in a variety of ways;
- learn and recite a poem.

Soapy Spuds

Dad was straining the potatoes
when he said to me:
Ssssh!
Don't tell anyone:
some of the spuds have fallen out of the colander
and into the sink.
I've put the ones that fell out
back in with the others.
No one'll notice.
Sssh!
Don't tell anyone.

So I didn't tell anyone.
And he was right.
No one noticed.
No one at all.
He got away with it.
Brilliant.

One problem:
my potatoes
tasted of washing up.

Thanks Dad.

Soapy Spuds

Dad was straining the potatoes
when he said to me:
Ssssh!
Don't tell anyone:
some of the spuds have fallen out of the colander
and into the sink.
I've put the ones that fell out
back in with the others.
No one'll notice.
Sssh!
Don't tell anyone.

So I didn't tell anyone.
And he was right.
No one noticed.
No one at all.
He got away with it.
Brilliant.

One problem:
my potatoes
tasted of washing up.

Thanks Dad.

"Soapy Spuds"

Discussion Points

- What did dad use to strain the potatoes?
- Why did he strain the potatoes?
- Why did the potatoes stay in the colander?
- How does a colander work?
- What else could you use to remove water? (sieve or filter)
- What else do you use a sieve or filter for?
- Is there a difference between sieving and straining? What is it?

Science Background

- Insoluble solids can be mixed with water then separated by leaving them to stand until the solids settle at the bottom of the container. Once the solid has settled at the bottom, the surface liquid can be poured off. This is called decanting.
- Sieves of various degrees of fineness can be used to separate solid-solid and solid-liquid mixtures. A filter is really only a very fine sieve used to separate solid-liquid mixtures.
- In industry, materials such as washed sand and gravel are also used as filters as in the case of water treatment in sewage works.
- Magnets are used to separate out metals, for example, in waste for recycling. Aluminium and brass are not magnetic and would not be attracted by a magnet.
- Solutions cannot be separated by any of these methods.

Key Ideas

- Solid particles of different sizes can be separated by sieving.
- Insoluble solids can be separated by decanting, sieving or filtering.
- There are other methods of separating insoluble solids, for example, using a magnet to separate magnetic metals.

Science Skills

Children should be able to :

- plan, sort, measure, observe and record data on a table;
- choose the appropriate apparatus and use it with care;
- work co-operatively with others.

Key Activities

Ask the children to sort a bucket of stones into different sizes using different size sieves. Challenge the children with a problem, such as how to get the rice out of a mixture of rice and dry peas.

Using some of the same solids as above, add water and get the children to recover the solids from the water. Demonstrate different ways of filtering coffee, making tea and using a strainer; look at tea bags.

Solids in water sink to the bottom. Even very muddy water will become clear if it is left to stand for a few hours. Look at different materials that might act as filters, for example: old nylon curtain, muslin, paper filters. Explain that filters are like sieves with very small holes. Use a microscope to look at the different materials.

With older children, look at filters used in industry such as sand or gravel for water purification.

Gravel cleans water well – the gravel must be clean before you start.

Make some muddy water and challenge the children to find out how clean thay can get it by using different filters. Do not allow children to drink the water.

Plastic lemonade bottles cut in half, with the tops turned upside down, and muslin to cover the open end work very well. Also, gravel, wire wool, cotton wool, coffee filters, sand are all good filters to test. Allow children to use and evaluate a range of filters.

Use magnets to separate out magnetic materials, for example, paper clips mixed with dry peas or paper clips hidden in sand.

Safety : Recognise risks and assess hazards. Use plastic gloves when handling soil. Take care with younger children when they are handling smaller particles such as peas or gravel.
See ASE publication *Be Safe!* for information on all aspects of safety in school science.

Numeracy Skills

Children should be able to :

- collect data;
- organise data;
- record data on a graph;
- produce a bar chart or a table;
- interpret data.

Literacy Skills

Children should be able to :

- use speech dialogue;
- use speech marks;
- manipulate words;
- understand the use of different words to express magnitude.

Drizzy Fink

Hail! Hail!
I come from another
galaxy
I have been learning English
I find some of your words
very hard to say.
I will now try to talk
about your fizzy drinks.
I like your fizzy drinks
I think I will drink
lots of fizzy drinks
and collect
lots of bottles
and lots of tottle bops
er
bopple tots
topple stobs pobble lots
no
stottle pobs
tobble spots
lottle slobs
lobble slops
Please
can you help me with this?
And please
can I have a drizzy fink?

Drizzy Fink

Hail! Hail!
I come from another
galaxy
I have been learning English
I find some of your words
very hard to say.
I will now try to talk
about your fizzy drinks.
I like your fizzy drinks.
I think I will drink
lots of fizzy drinks
and collect
lots of bottles
and lots of tottle bops
er
bopple tots
topple stobs pobble lots
no
stottle pobs
tobble spots
lottle slobs
lobble slops
Please
can you help me with this?
And please
can I have a drizzy fink?

"Drizzy Fink"

Discussion Points

It would be useful to have a bottle of lemonade to look at when discussing this poem.

- If I shake the lemonade bottle, what do you see?
- What do you think makes some drinks fizzy?
- What do you think the bubbles are in fizzy drinks?
- If I take the top off the bottle what do you hear?
- What is in the lemonade that causes this to happen?
- Is fizzy lemonade just a liquid?
- What else is in it?
- Why do you think that fizzy drinks eventually go "flat"?
- What happens to the fizz?
- Do you think we can measure how much of the mixture is fizz?
- How do you think the gas could be separated from the liquid?

Science Background

- A mixture is two or more substances physically but not chemically combined.
- Mixtures can be: solid in solid, solid in gas (smoke), liquid in liquid (emulsion – milk) liquid in gas (aerosol or clouds, mist, and fog), gas in liquid (fizzy drinks), solid in liquid (suspension – muddy water), solid in liquid (solution – salt water). You will need to explain the difference between suspension and solution.
- In general, solutions cannot be separated by physical means but there are some that can be separated physically, for example, by: sieving, filtering, evaporating etc. This may be a difficult process, as with some gases.
- Mixing materials may cause changes that are permanent such as with plaster of Paris mixed with water. This is a chemical change that forms a new material; it is not a mixture because the original materials cannot be recovered. Older children need to be made aware that there are variety of mixes and various different reactions.

Key Ideas

- Mixtures have more than one thing in them.
- Mixtures are physically, not chemically, combined.
- They can be separated.
- Mixing different substances can cause permanent changes.
- To become familiar with the property of a gas; it expands as the temperature rises.
- To know that carbon dioxide is a gas.

Science Skills

Children should be able to :

- plan a fair test;
- make reasoned predictions;
- choose appropriate apparatus and use it properly;
- make accurate observations and measurements;
- record results accurately;
- communicate results to others;
- work with others.

Key Activities

With the children, look at and discuss obvious mixtures, for example: Dolly Mixtures, frozen mixed vegetables, jar of mincemeat, or a breakfast cereal such as muesli. What do the mixtures contain?

Then look at mixtures in which the ingredients are not obvious, such as lemonade or a cup of coffee. Try to find out the contents of the mixtures. Could these mixtures be separated? If so, how? Younger children could design and make their own mixed breakfast cereal or fruit salad. Older children could design a salad dressing. Older children could also investigate and measure the ingredients of a sample of mixture. They could write a recipe for others to follow.

Fizzy drinks are a mixture of gas and liquid. Take a range of fizzy and non-fizzy drinks into the classroom and encourage children to use all their senses to describe the differences between fizzy and non-fizzy drinks, for example:

Do they look any different when they have been left to stand for a while?

What happens when they are shaken?

What sounds do they hear when a fizzy or a non-fizzy drink is opened?

Why do they think one makes more sound than the other?

Does a fizzy drink taste different to a non-fizzy drink?

Drop currants into fizzy lemonade. Observe what happens. Why do the currants go up and down?

Encourage children to think about how the bubbles (gas) got into the fizzy drink. Ask them to list their ideas. Then ask them how they think they could take the gas out of the drink. The following explanation is from R. Feasey and B. Gallear, *Primary Science and Numeracy*, ASE 2000.

1. Weigh a full lemonade bottle with its lid on and record the results.

2. Replace the lid with a bung and tubing which goes into a gas measuring cylinder and has a Hoffman screw clip to act as a valve to let the gas through.

3. Ask someone to shake the bottle while another child holds the cylinder and the control clip.

4. Open the Hoffman clip to allow the gas through and displace the water in the measuring cylinder.

5. Measure the amount of gas and compare it with the stated amount of lemonade in the bottle.

6. Weigh the bottle and calculate the difference.

7. This procedure may need to be repeated several times to remove all the gas from the lemonade bottle. In fact, from some one litre bottles around three litres of gas can be extracted.

Safety : Recognising hazards and controlling risks. Handling hot and cold things. Handling glassware. Do not allow young children to taste food.
See ASE publication *Be Safe!* for information on all aspects of safety in school science.

Numeracy Skills

Children should be able to :

- calculate the gas in a bottle of pop;
- measure and record temperature;
- produce a table of results.

Literacy Skills

Children should be able to :

- manipulate words;
- measure words;
- use rhyming and rhythm.

Making Jelly

It's my job
to take the slab of jelly
and break it up into cubes.

It's mum's job
to pour on the boiling water
to melt the cubes.

It's my job
to stir it up
until there are no lumps left.

It's her job
to put the bowl
in the fridge to help it set.

It's my job
to eat it.

5 it's my job to eat it.

"Making Jelly"

Discussion Points

- Can you describe how to make a jelly?
- What happens when you add water to the jelly?
- "...To melt the cubes"– is the poet correct when he uses this word? Why?
- What is the difference between melting and dissolving?
- How does the jelly change when you add water?
- What other solutions can you think of that change when water is added?
- Why was boiling water used, could you use cold water?
- Why was the jelly stirred?
- Why was the jelly broken up?

Science Background

- When a solid is added to a liquid, if the resultant mixture is clear then the solid has dissolved and formed a solution. This solution cannot be separated by filtering even though no new chemical substance has been made. It can only be separated by evaporation. If the result is a cloudy mixture then the solution is a suspension which will eventually settle out and can be separated by filtering.
- Any substance which dissolves is said to be soluble and there are various factors that affect the dissolving: temperature, size of solid particles, stirring, amount of solid (solute) added. Varying amounts of solute can be added to the solvent (liquid) but there is a limit known as the saturation point. The saturation point varies with the substance and is also affected by temperature. Normally, more solid will dissolve in hot water than in cold water. This is why when a hot, saturated solution cools down solid crystals may form.
- Liquids other than water can be used to dissolve solids that will not dissolve in water. However many of these solvents are harmful and cannot be used by children in the primary school. Some may be demonstrated by the teacher, for example, showing that nail varnish dissolves in nail varnish remover.

Key Ideas

- Some solids dissolve in water to give solutions, others do not.
- Various factors affect dissolving, for example, stirring or particle size or temperature.
- There is a limit to the mass of solid that can dissolve in water. This limit is different for different solids and at different temperatures.
- Substances that do not dissolve in water can sometimes be dissolved by other solvents.

Science Skills

Children should be able to :

- plan a fair test;
- use a thermometer and stopwatch accurately;
- make accurate observations and record results on a table;
- work co-operatively in pairs or in a goup.

Key Activities

Provide the children with a variety of everyday soluble and insoluble materials such as: salt, sugar, coffee, flour, sand, to investigate solubility. Ask children which materials dissolve.

Repeat the above for some of the soluble solids using different amounts of water. Time how long each takes to dissolve. Ask the children to record their results. Discuss with them what effect different amounts of water and different temperatures will have when dissolving soluble solids.

Get them to repeat the test this time stirring the liquid as the solid dissolves. Again record the times.

Repeat the test again using one solid with different-sized particles such as sugar cubes, sugar crystals, granulated sugar, and icing sugar.

Investigate the fastest way to make a jelly. It may be better to do this as a teacher demonstration, as, depending on their cultural background, not all children will have experienced jelly making.
There are many variable factors here affecting the result. Young children will need help deciding what they are, for example, the temperature of the water or the sizes of the pieces of jelly.

Investigate if there is a limit to the amount of sugar that will dissolve in $100cm^3$ of water. Find out if this is the same for salt.

Then try using hot water and cold water. Does the temperature of the water make any difference? A line graph can be constructed of the results.

Demonstrate propanone (acetone) dissolving nail varnish and white spirit dissolving cooking oil. Show the difference by trying to dissolve them in water. This is a good demonstration activity to introduce children to other solvents.

Safety : Take care when using hot water. Use spirit thermometers. Children must not use hazardous solvents. Beware of asthmatics breathing flour dust. See ASE publication *Be Safe!* for information on all aspects of safety in school science.

Numeracy Skills

Children should be able to :

- choose and use the appropriate equipment for measuring capacity;
- estimate capacity;
- measure volume;
- record volume;
- read the scale on a measuring cylinder.

Literacy Skills

Children should be able to :

- understand and use possessive nouns;
- read aloud;
- compare views;
- write instructions for making a jelly.

Steamy Shower

I love a
dreamy, steamy shower
hanging about for over an hour
just before bed
getting hot and red
in the steam
standing there with time to dream
water-running-over-me feeling
drips dripping off the ceiling
mum says its my fault its peeling
nothing can beat
the hot wet heat
nothing wetter
nothing better
I love a
dreamy steamy shower

Steamy Shower

I love a
dreamy, steamy shower
anging about for over an hour
just before bed
getting hot and red
in the steam
standing there with time to dream
water-running-over-me feeling
drips dripping off the ceiling
mum says it's my fault it's peeling
nothing can beat
the hot wet heat
nothing wetter
nothing better
I love a
dreamy steamy shower

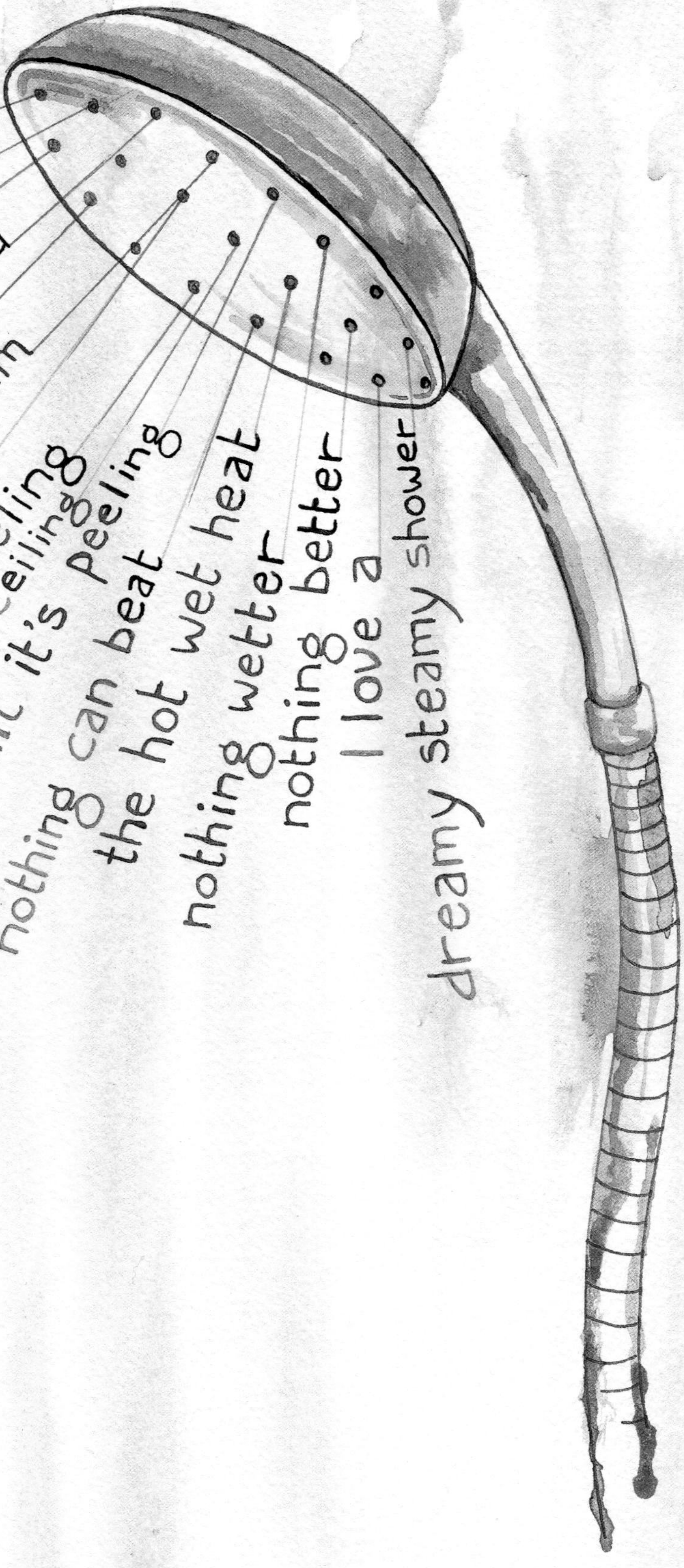

"Steamy Shower"

Discussion Points

- What gases are in the bathroom?
- Why is water vapour there? Where does it come from? Can you see it? What shape is it?
- What shape is the water in the bath? Does it stay like that?
- When water becomes warmer it changes more quickly from a liquid to a gas (vapour). This is called evaporation. What will eventually happen to it?
- What is the water dripping from the ceiling? Where does it come from?
- There is a water-cycle happening in the bathroom. Can you describe it?
- How would you describe steam? Where does it come from?
- When it rains, there are puddles on the playground. When the puddles dry up, where does the water go?
- Where does the water go when washing dries?
- Do puddles and washing dry up quicker on a hot or cold day?
- Describe a water cycle in a kitchen.

Science Background

- A solid has a definite shape that remains the same unless a force is acting upon it. The particles that make up a solid are close together and move about a fixed position. A liquid has no fixed shape but a fixed volume and takes on the shape of the container it is in. The particles remain in close contact with each other but have more energy and movement than a solid. A gas has no fixed shape or volume. Energy, usually in the form of heat is required to change from solid to liquid to gas. This process is reversible, which happens on cooling and a transfer of energy.
- Steam is the gaseous state of water at or above 100ºC; it is not visible. Water vapour is the gaseous state below 100ºC; it is also usually not visible.
- Evaporation occurs when a liquid dries up. Those molecules on the surface of the liquid which have more energy change to a colourless gas called water vapour. They move from the liquid into the air. The more liquid that is exposed to the air, the faster it will evaporate.
- Evaporation occurs at low temperatures but increases as the temperature rises because the surface molecules have more energy and move away faster. As evaporation occurs it has a cooling effect.
- When water is boiled, the gas phase is very hot (100ºC). When the water vapour or steam hits colder air, the gaseous state turns back to the liquid state (condenses) and forms tiny droplets in the air which remain suspended as a fog, mist or cloud. If these tiny droplets hit a cold surface they coalesce further to form water.

Key Ideas

- Water changes from a liquid to a gas called water vapour, which is in the air.
- Evaporation is also affected by the size of the surface area of liquid from which the evaporation is occurring.
- Evaporation is reversed when the water vapour cools. This is called condensation.
- Solids that have dissolved can be recovered by evaporating the liquid from the solution.

Science Skills

Children should be able to:

- follow instructions correctly;
- use apparatus carefully;
- plan and carry out an investigation;
- work with others.

Key Activities

Ask children to think about what happens in the bathroom when someone has a bath or shower. They could create a flow diagram or draw pictures and annotate them.

Discuss the children's ideas and then do a teacher demonstration, using very hot water, to illustrate how water vapour condenses when it meets a cold surface.

Talk about evaporation and condensation with the children. Encourage them to think this through in the context of the bathroom and kitchen.Then help them to make links with other contexts where evaporation takes place such as washing on the line or puddles on the ground.

Children could be given the opportunity to investigate what makes a puddle dry up. They could create their own puddles by pouring water on different parts of the school playground, using the same amount of water but in different places, such as shady or not shady.

This could be done each day of the week, noting the weather conditions. Do they find a link between the weather and how quickly or slowly the puddle dries up?

Safety : Take care when using thermometers, and handling hot and cold liquids.
See ASE publication *Be Safe!* for information on all aspects of safety in school science.

Numeracy Skills

Children should be able to:

- use a stopwatch accurately;
- make careful observations;
- measure area accurately;
- record results accurately;
- use a thermometer.

Literacy Skills

Children should be able to:

- use rhyme and alliteration;
- manipulate words;
- make a flow diagram to explain something;
- read poetry and modify performance.

Cow's Shoes

If a field is boggy
a cow's feet go soggy
what's more it's feet get sore.

Now here's some news
they've invented cow shoes
if you go into a shoe shop now
you might bump into a cow.

ow's shoes

If a field is boggy

a cow's feet go soggy

what's more it's feet get sore

Now here's some news

they've invented cow shoes

if you go into a shoe shop now

you might bump into a cow

"Cow's Shoes"

Discussion Points

- What animal is the poem about?
- What is the news in the poem?
- Has anyone walked through a cow field? What is the ground like during the winter?
- What do you wear on your feet in a muddy field and why?
- What do cows' feet look like? How are they different to our own? Do they wear shoes?
- What problem do the cows have because of the boggy field?
- What do you think is the job of the cow shoe?
- Which other animals wear shoes?
- What do they look like? What job do they do?
- Do you think a cow would care what his shoes felt like or looked like?
- What do you think is good about the shoes you are wearing today?
- What is particularly good about your shoes (extra grippy, trendy, cool)?
- What do you think cow shoes would look like?

Science Background

- The idea of cows' shoes is based on a Millennium design. It sounds far fetched but it does have some practical applications. Humans wear shoes to help keep feet warm and dry and to protect them. Other animals do not have this luxury. For example, a cow's feet are subject to all weathers, hard, or wet and muddy surfaces, and infection. The feet of animals take the weight of the whole body and in the lifetime of an animal take some pounding.
- Animals and birds have feet to suit their environment and way of living. For example, some birds have talons for catching prey and tearing, others have feet that are able to grasp round branches. Looking at the feet of different creatures can tell a lot about their habitat and feeding habits.
- Humans' shoes are a product of science: the types of materials, looking at people's feet, and considering some of the problems many people suffer in terms of posture. Some people have difficulties because of ill-fitting shoes and fashion trends that not only cause problems with feet but also hip and back problems.

Key Ideas

- Products are made for a particular purpose.
- Certain materials are used to do certain jobs.
- Objects are made of component parts joined together. Each component has a purpose.

Science Skills

Children should be able to :

- plan and carry out a fair test;
- work with a partner.

Key Activities

Set up a shoe shop for the children to explore what shoes look like, what they are made from, why they are worn, favourite shoes and comfort and suitability.

Ask them to look at the soles and uppers of each shoe. What materials are used? Why do they think those materials were chosen? Match these to named samples. How are the shoes kept on? Learn the names given to the component parts of shoes (sole, upper, heel, tongue). What kind of material is each part made from and why?

Disassemble a simple flip-flop or shoe. Look at the components – which materials are used and how were the different parts joined? Ask the children to draw simple diagrams of the parts of the shoe and to label the parts.

Get the children to work in pairs. Ask them to use a piece of card to make the sole of a shoe. Demonstrate cutting round the shape of a foot, leaving a margin around the foot to cut out "sole" from card. Repeat the task using fabric instead of card. Explore the suitability of different scissors on different materials.

Explore techniques for joining materials to themselves and each other (thick card, layered card, or polystyrene tiles for example) using glues, staples, holes and knotting methods, or sewing with a simple running stitch. Which methods are best with which materials? Which methods are used on the sandals you looked at?

Challenge the children to design and make a summer beach sandal. Explain that the task is to design a shoe that could be worn on a sandy beach.

They will have to think about the following:

Who will the shoe be for?

What size will it be?

What will it be made from?

Draw a design on paper reminding the children of work done previously. Discuss how they will make their sandal.

Ask them what material they will use for their sandal and what tools will be best for the task. Get them to make the sandal. Carry out stages focusing on the methods that have been explored and the use of appropriate tools.

When they have finished find out if they are happy with what they have made? Is it what they wanted? Does it fit the person it is made for?

This activity could be extended by asking the children to design footwear for other creatures, for example, elephant "wellies" or polar bear "slippers".

Safety : When carrying out a risk assessment of this activity, attention must be paid to appropriate tools, storage, and adult supervision.
See ASE publication *Be Safe!* for information on all aspects of safety in school science.

Numeracy Skills

Children should be able to :

- identify patterns in data, for example, shoe size related to height;
- make appropriate measurements;
- do a survey of shoe types worn in school.

Literacy Skills

Children should be able to :

- identify patterns of rhythm and rhyme;
- realise when poetry is humorous;
- make a flow chart of how a shoe is made;
- make a poster advertising shoes.

Incredible Shrinking Car

I am
The Incredible Shrinking Car.
I started out my life
as a giant chunk of steel and chrome
cruising round town
gobbling up the geography
fighting to find a place big enough
in the Supermarket Car Park.

Bit by bit I have shrunk.
I've lost my mighty bumpers.
That overhanging cliff, we call, "my boot"
has been chopped off.
My bonnet's been squeezed
towards the windscreen
and the gate-sized doors
are now no more than little hinged windows.

My engine
which once roared its head off
now mutters to itself
and I slip into corners
like 20p goes into the slot
of a parking meter.
My only worry is: when will it stop?

Will I end up as a
Matchbox toy
on the bedroom floor
of the boy
next door...?

Incredible Shrinking Car

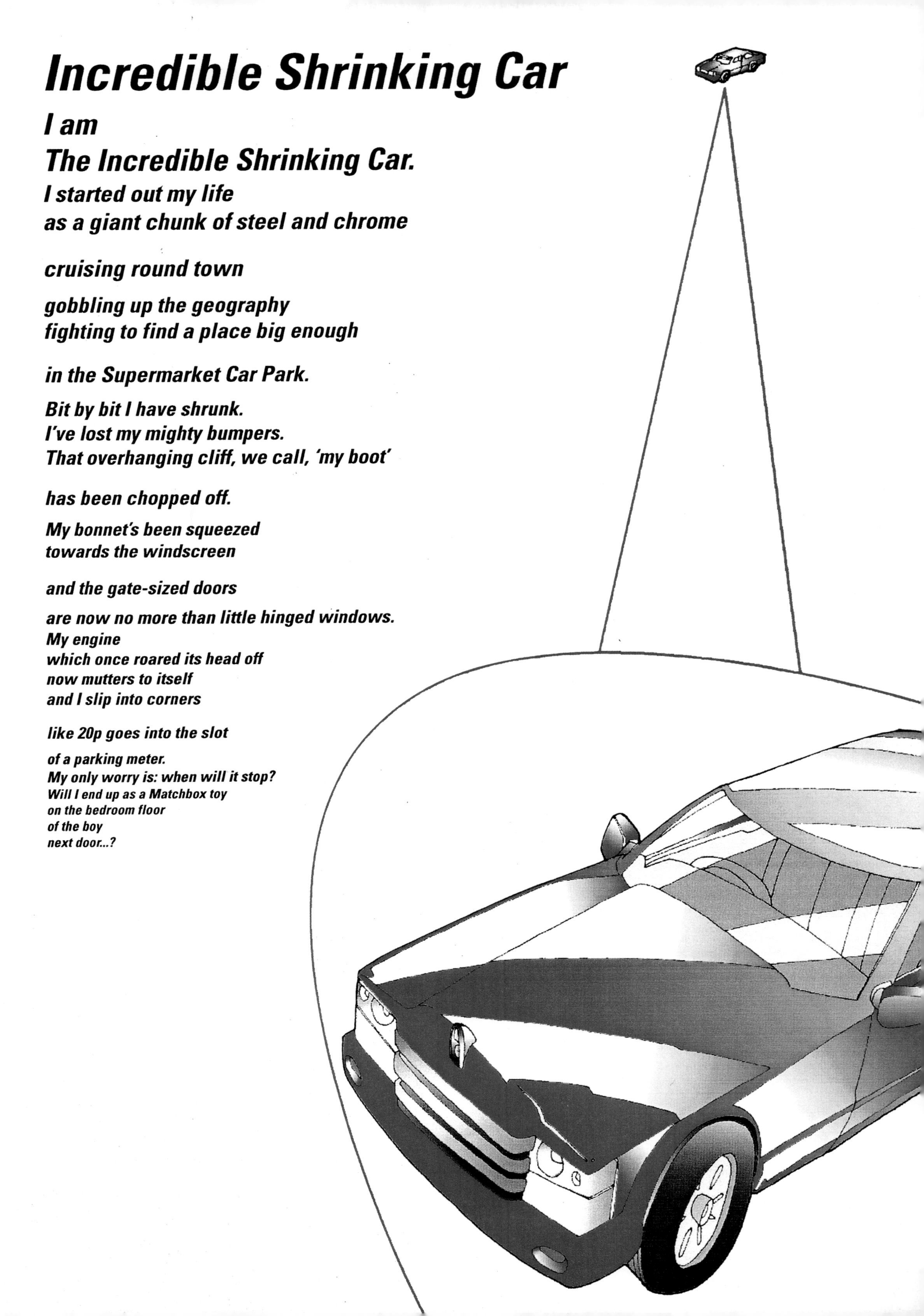

I am
The Incredible Shrinking Car.
I started out my life
as a giant chunk of steel and chrome

cruising round town

gobbling up the geography
fighting to find a place big enough

in the Supermarket Car Park.

Bit by bit I have shrunk.
I've lost my mighty bumpers.
That overhanging cliff, we call, 'my boot'

has been chopped off.

My bonnet's been squeezed
towards the windscreen

and the gate-sized doors

are now no more than little hinged windows.
My engine
which once roared its head off
now mutters to itself
and I slip into corners

like 20p goes into the slot

of a parking meter.
My only worry is: when will it stop?
Will I end up as a Matchbox toy
on the bedroom floor
of the boy
next door...?

"Incredible Shrinking Car"

Discussion Points

An advertisement for a small car would be helpful for use after reading this poem.

- What has happened to the car in the story? What has been "chopped off" the car? What has been "squeezed" onto the car?
- The poem says the engine used to roar. What does it do now? Why do you think this is?
- What does the car tell us it is particularly good at now?
- What does the car think might happen to it next and why?
- What different jobs are there to be done in the manufacture of a car?
- Why do you think car manufacturers are spending large amounts of money on producing smaller cars? What are the benefits?
- How small do you think a small car could become? What limitations are there on a car becoming smaller and smaller? What disadvantages do small cars have? Who would find them most useful and for what purposes?
- In the poem we hear about changes to the outside of the car but what changes do you think happen to the inside of the car when it becomes smaller?
- Are there any ways in which cars can be made more environmentally friendly?

Science Background

- Cars are made of a large number of component parts all of which have specific names and purposes. The components are linked together. Consider how a working mechanism has been achieved
- Manufacturers now produce small cars designed to drive with all the styling, sophistication and features associated with larger cars.
- The numbers of vehicles on the road is increasing all the time and every car, lorry or van etc. takes up space and adds to pollution.
- The incredible shrinking car may not be far off in terms of science and technology because scientists are exploring ways to make smaller, more environmentally friendly cars.
- Such approaches mean looking at different materials and fuels for cars. New fuels, such as hydrogen gas, are much cleaner than petrol; other cars use electric batteries which can be re-charged after a journey. At present electric (battery powered) cars cannot reach the speed of petrol cars and the batteries limit the distance that can be travelled.
- However some cars now being developed are hybrids which use both a petrol engine and a battery. Computers inside the car are able to switch between the two systems to power the car.
- Other types of ways to power cars include solar energy and biogas; such cars could be driven in the future particularly as fossil fuels are not renewable.

Key Ideas

- Technology to control the environment.
- Technology and demand for resources.
- Technology responding to values and scientific progress.
- Design and manufacturing processes.
- Selecting and using design processes.
- Properties of materials in relation to their practical use.

Science Skills

Children should be able to :

- research a topic;
- collect information and present it to an audience;
- be able to work cooperatively with others.

Key Activities

Discuss with children some of the issues related to road vehicles. Make a list of the advantages and disadvantages of vehicles and also talk about alternative methods of transport, for example, trains, bicycles, walking, car sharing.

Give children time to research car manufacture and alternative ways of powering cars such as electric batteries or solar power. Children could produce a table indicating the advantages and limitations of each of the alternative energy sources.

Ask children to think about the future and the problems of having more and more cars on the road. What do they think the car of the future should look and be like? Then ask them to design the car of the future and to annotate it, indicating its special features with regards to power, materials etc.

Children could be challenged to design cars for different purposes, for example, short journeys less than 20 miles and long haul journeys.

Ask children to indicate how their car is environmentally friendly.

What special features does it have that make it environmentally friendly?

What is its fuel consumption?

How much pollution does it create?

Can it be recycled?

Finally ask children to produce a manufacturer's brochure, poster advertisement or television advertisement to sell their car to the public who are looking for environmentally friendly transport.

Safety : Take care when using scissors and glue. See ASE publication *Be Safe!* for information on all aspects of safety in school science.

Numeracy Skills

Children should be able to :

- ◆ read and compare specifications for different cars, for example, speed, fuel consumption etc.
- ◆ make a table of the comparisons;
- ◆ produce a bar chart from the data.

Literacy Skills

Children should be able to :

- ◆ use appropriate vocabulary;
- ◆ use presuasive writing to support an argument;
- ◆ produce a brochure for a small car;
- ◆ create a television advertisement;
- ◆ perform a role play, for example, interview the car's designer.

My Supermarket Queue

When I go shopping at the
supermarket
I have a special queue of my own.

It's the one where
the man in front
has had a massive pile of stuff
all checked out
and he's packed all his bags
and he's discovered he hasn't got
enough money.
So he starts taking out stuff
he doesn't think he needs
and he holds it up
one by one
and he says,
"How much is this?"
and he's trying to work out
what he can leave
and it all takes hours
in my own special queue.

It's the one where
the woman in front
bought an avocado
and it's lost its price tag
and avocados don't have bar codes
and the checkout guy
doesn't know the price
and the suprevisor
doesn't know the price
and when she goes off
to find the price
the price tag has
fallen off there too
and it all takes hours
in my own special queue.

And it's the one where
the bag of onions splits
and bar code doesn't beep
the till-roll jams
the conveyor belt stops
the trainee presses CANCEL
and has to start all over again
while a baby screams
like an aeroplane
because his brother
nicked his biscuit

and it all happens
in my own special queue
and
it
takes

<u>hours!</u>

My Supermarket Queue

When I go shopping at the
supermarket
I have a special queue of my own.

It's the one where
the man in front
has had a massive pile of stuff
all checked out
and he's packed all his bags
and he's discovered he hasn't got enough money.
So he starts taking out stuff
he doesn't think he needs
and he holds it up
one by one
and he says,
'How much is this?'
and he's trying to work out
what he can leave
and it all takes hours
in my own special queue.

It's the one where
the woman in front
bought an avocado
and it's lost its price tag
and avocados don't have bar codes
and the checkout guy
doesn't know the price
and the supervisor
doesn't know the price
and when she goes off
to find the price
the price tag has
fallen off there too
and it all takes hours
in my own special queue.

And it's the one where
the bag of onions splits
a bar code doesn't beep
the till-roll jams
the conveyor belt stops
the trainee presses CANCEL
and has to start all over again
while a baby screams
like an aeroplane
because his brother
nicked his biscuit

and it all happens
in my own special queue
and
it
takes

hours!

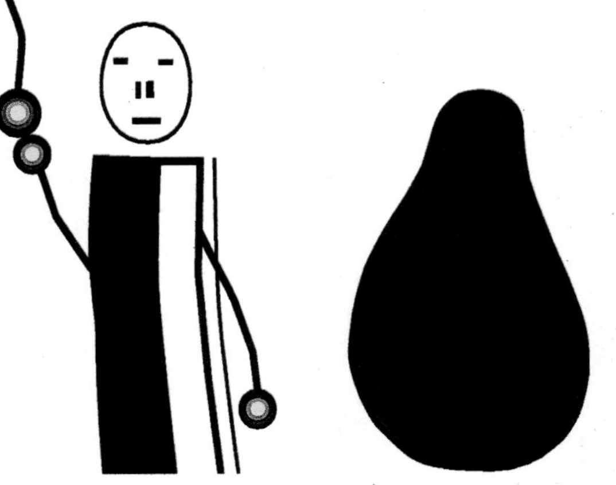

5 001999 002000 > CANCEL

"My Supermarket Queue"

Discussion Points

- How often do you go to a supermarket and what do you buy?
- What problem is the person in the poem having?
- Why did the "man in front" have to start unpacking his bags?
- What is wrong with the avocado?
- What is a bar code and how does it work? What does it do other than tell the price?
- Do all shops use bar codes or do some just ring up the price?
- What other problems are there in the writer's special queue?
- What systems do supermarkets use to try to help shoppers through the checkout quickly?
- Do all supermarkets have the same place for checkouts, or the same systems operating?
- What problems do you think could arise from the invention of this new product?
- Are there any other systems used in a supermarket to help the shopper?
- What systems are used in school to help people around? These may include: dinner times, going to assembly, going in at breaks, school shop, use of the playground space or fire drill. What do you think of them?

Science Background

- Science and technology permeate all aspects of daily life. For example, laser technology and computer technology help make shopping quicker, by scanning your purchases at the till.
- Bar codes are commonly used in supermarkets today. At the till a scanner uses a laser beam to read the bar code on each item; the beam is reflected from the bars on the outside of the product in pulses which are changed into electrical signals. The bar code tells the central computer in the shop what the item is, for example, a bag of sugar, and the computer finds out the cost of the sugar. The computer then passes the information to the till where the name and cost of the item are printed on the receipt. This is a very quick and efficient way of automatically finding out and recording the price of each item. However, if the bar code is crumpled or greasy the laser will not always be able to read the information.
- Laser and computer technology enables supermarkets to change information of thousands of products quickly without having to re-price every item individually with a price sticker.
- In addition, by reducing stock levels held on the computer each time an item is scanned, the store has a continuously up-to-date record of stock. This information is used for re-ordering purposes.

Key Ideas

- Systems can be recorded and communicated.
- Systems can be evaluated using observation and surveys.
- New systems can be tested by modelling.

Science Skills

Children should be able to :

- work in groups;
- plan and carry out a practical task;
- make observations and record results;
- use different methods to communicate ideas and explain findings.

Key Activities

Get the children to identify a system within the school context. Organise the class into groups and ask each group to brainstorm the last question from the discussion points. Each group decides which issue is most cause for concern, then records the existing system (maybe using a plan of the school or areas) and notes any of the problems they feel arise from this.

Look at existing school systems and how they are recorded, such as maps used in playgrounds for play areas, written instructions to teachers, litter areas, childrens' notices. Which are easiest to understand – look for use of colour coding, area demarcation, instructions and keys. Which are most likely to be adhered to?

Give the class a system to evaluate such as the use of playground. Together disassemble the system.

Who uses it (children or adults involved and their present role, such as class teachers or dinner duty staff)?

Where does it take place?

Why is the system in place?

What are its advantages and disadvantages?

Discuss the environment this takes place in?

Is it adequate?

Is it a cause of a problem?

Could the environment be realistically changed?

Do the people using the system affect it by their behaviour or numbers?

What parts of the system are successful? Why?

What parts are unsuccessful? Why?

Again, working in groups, ask them to design a new or improved system for their selected school system, for example, a new school fire drill.

Get them to make a design on paper or to make a model of their new system to test the effectiveness of their ideas. For example, are there "bottlenecks" or danger points? What new equipment may be needed (chalk lines, class cards to hold up in the playground, or number cards)? What will they need to look like? What could they be made from?

Ask each group to present their new system to the rest of the class. How will it be best explained, with a poster or an OHP? Discuss class responses and questions arising.

If possible do dummy runs of the new systems and evaluate responses and problems. Maybe these could be presented to a school council or headteacher for consideration.

Safety :
See ASE publication *Be Safe!* for information on all aspects of safety in school science.

Numeracy Skills

Children should be able to :

- recognise coins or notes of different values;
- use knowledge of number bonds to calculate mentally the change needed when shopping;
- carry out a survey.

Literacy Skills

Children should be able to :

- use the structure of a poem and substitute a verse using their own words and ideas;
- read a poem by a favourite poet and compare their work;
- take part in a discussion and make an oral presentation;
- use a flow diagram to represent a system;
- make connectives.

Night Rides

...and when the house
is quiet
and everyone sleeps
sometimes I wake up
and breathe in darkness
for so long
I have to get up
open the curtains
and look out over the city
watching the cars do U-turns
that they don't dare do
in the day.
a cat talking to a bin
and a bus full of lovers
sailing down the High Street
lit up like a fairground.

...and I think of my big sister
far away
and I want to ride that bus
all the way
up to her block of flats
up the stairs
and in through the door.

The city winks.
I'm cold
...and the bus sails on.

.and when the house
; quiet
nd everyone sleeps
ometimes I wake up
nd breathe in darkness
or so long
have to get up
pen the curtains
nd look out over the city
atching the cars do U-turns
nat they don't dare do
the day,
cat talking to a bin
nd a bus full of lovers
ailing down the High Street
up like a fairground.

.and I think of my big sister
ar away
nd I want to ride that bus
l the way
to her block of flats
the stairs
nd in through the door.

he city winks.
m cold
.and the bus sails on.

Night rides

"Night Rides"

Discussion Points

Note: The language of this poem may be difficult for very young children.

- What time of day do you think it is in the poem?
- How could the city "wink"?
- Have you ever felt that you could "breathe in darkness"? What was it like?
- What can you see from your bedroom if you look out at night?
- Why/how could the person in the poem see the cars, the cat and the bus? What else might he or she be able to see?
- Why is the bus described as "sailing down the High Street" and "lit up like a fairground"?
- What could you use to make the side of a bus light up?
- What do you usually see on the back and sides of buses?
- Why do people pay bus companies to put their adverts on their buses?
- What is the advantage of the advert being illuminated at night?
- Have you ever seen things advertised on buses that are for a charity or campaign?
- Which colours are best for advertising?

Science Background

- Light travels very fast – "at the speed of light".
- It travels in straight lines, radiating in all directions from its source.
- Sources of light can be primary, that is direct, for example, from the Sun or flames from a fire, or secondary, that is indirect, for example, from the Moon or light bulbs.
- There are many sources of light in everyday life but most are indirect, especially in the day-time when nearly all the direct light comes from the sun.
- We need light to see things. Light from an object, usually scattered or reflected, enters the eye through the pupil. It reaches the retina at the back of the eyeball which sends electrical signals to the brain. These electical signals describe to the brain the object that is being "seen".
- Darkness is the absence of light, and shadows are caused when light is completely or, more usually, partially blocked by opaque materials which do not allow light to pass through them; shadows are thus regions of darkness. Transparent and translucent materials, such as glass, transmit light: that is, they allow light to pass through them.

Key Ideas

- Light travels from a source.
- There are different sources of light.
- Dark is the absence of light.
- We see when light enters our eye. The light travels from the object to our eyes.

Science Skills

Children should be able to :

- follow instructions and make a work-plan;
- work with others.

Key Activities

Ask children to imagine looking at a town or city at night. Ask them which are direct/primary sources, that is, they produce their own light, and which are indirect/secondary sources, that is, they scatter or reflect light. Then ask the same questions for the same scene during the day. Be sure to mention the Sun, Moon and stars. Get them to draw a table showing the light source and whether it is primary or secondary.

For younger children start with the classroom or a room at home and tell them to draw a picture of the sources of light. Where is the light and how does it get to us? Ask children what is the difference between light and dark. Try to bring out the idea that dark is the absence of light.

Use a torch to illuminate an object in the classroom, and ask children to explain and/or draw how they see it. What is happening to the light? Then ask the same question about an object which is easy to see and well-lit, but is much further away, say about 500m. This is to establish that light travels and that we see when light travels from an object to our eyes.

Older children can be challenged to design and make an illuminated display to advertise a good cause. Mount it on a frame to attach to a model or picture of a bus. Make a large card cut-out to give the children an idea of scale. The tasks will involve planning, designing, making and evaluating.

The children will have to decide what is going to be made and how.

They will need to:

generate ideas and to develop a clear idea of what has to be done;

consider the users and the end-purpose of the article;

develop criteria for their designs;

clarify their ideas and suggest ways forward;

consider appearance, function, safety and reliability;

propose a sequence of actions, and suggest alternative methods if things go wrong;

evaluate their design ideas and indicate improvements;

select materials, tools and techniques;

plan the steps to be taken to make the article;

evaluate the finished product, identifying strengths and weaknesses, and carrying out appropriate tests.

Throughout the making procedure encourage the monitoring of ideas and their practicality.

Discuss what they have made in terms of appearance, materials, colour, size or shape and fitness for purpose. Suggest improvements.

Safety : Never look directly at the Sun without protective eyewear. Make sure the appropriate battery and bulb voltage is used. Do not use re-chargeable batteries. Take care when connecting batteries.
See ASE publication *Be Safe!* for information on all aspects of safety in school science.

Numeracy Skills

Children should be able to :

- read bus timetables;
- estimate journey times;
- calculate journey times.

Literacy Skills

Children should be able to :

- understand and use verbs;
- understand the use of metaphore and simile;
- understand how to use words to change mood;
- use the appropriate vocabulary for naming and describing equipment, materials, components and processes;
- link clauses.

Periscope

The boy next door
is a bit of a dope.

I'm very clever.
Not like him.
I hope.

Between our houses
there runs a wall.
The wall is tall.
But I am small.

D'you know this
means: because
I'm small...

...I can't see over
that wall.

There's nothing
that I'd like more
than to look over that
wall at him next door.

P'raps when I'm
older I'll be tall.

Then I'll be able
to look over that wall.

MEANWHILE...

for my Birthday
I very much hope
I will get a
periscope.

Periscope

The boy next door
is a bit of a dope.

I'm very clever.
Not like him.
I hope.

Between our houses
there runs a wall.
The wall is tall.
But I am small.

D'you know this
means: because
I'm small...

...I can't see over
that wall.

There's nothing
that I'd like more
than to look over that
wall at him next door.

P'raps when I'm
older I'll be tall.

Then I'll be able
to look over that wall.

MEANWHILE...

for my Birthday
I very much hope
I will get a
periscope.

"Periscope"

Discussion Points

- What is a periscope?
- Why does the person in the poem want one?
- What does it look like?
- How big is it?
- Have you ever seen or used one?
- Where was it and and why was it there?
- What does it do?
- How do you think a periscope works?
- Why does a periscope have mirrors?
- Why are they in special positions?

Science Background

- All surfaces scatter light in many different directions. Smooth, shiny surfaces such as mirrors, both plane (flat) and curved, reflect light in a symmetrical and predictable way. Therefore the direction in which the light is travelling is changed in the same symmetrical and predictable way. When light reflected from an object enters our eye, we call this the image and we see a reflection.
- When light is reflected off two plane mirrors in succession, the laws of reflection are still obeyed and different effects are produced depending on the distance and angle between the mirrors. In a periscope the mirrors are separated, parallel and at 45° to the light path. This produces a shift in the path of the light, vertically in this case, but still allows the light to continue in the same direction.

Key Ideas

- Light is reflected by a smooth, shiny surface in a regular way.
- When light from an object is reflected into our eye by a mirror we see a reflection of the object.
- An object between two mirrors can produce more than one reflection of itself.

Science Skills

Children should be able to :

- collect and record data;
- use data to record patterns;
- use data to make generalisations and draw conclusions.

Key Activities

Younger children can be given a variety of reflective and non-reflective materials for them to investigate. Give them a torch to shine at the different materials. Ask them to find out which materials reflect and which do not and which is the best reflector of light. Ask them if the best reflectors of light give the best reflection of themselves?

On a large sheet of paper, using a ray box or a powerful torch masked to produce a narrow beam, demonstrate how a beam of light is reflected symmetrically like a ball bouncing off a flat surface – by a plane mirror placed at different angles to the beam. Allow the children to explore the effect.

Allow older children to include curved mirrors – plastic safety mirrors can be bent on the spot – two plane mirrors edge-to-edge at acute angles up to 90 degrees (as in a kaleidoscope), and two plane mirrors, spaced apart, at 45 degrees to the light path and parallel to each other (as in a periscope). Again, look for how the direction of the light path is changed. Record the activities by drawing the positions of the mirrors and beams of light. Ask the children to record the patterns that can be seen.

Give the children, in pairs, a plane mirror and ask them to explore how they can use the mirror to see a number of different objects in the classoom. Ask them to draw the positions of the mirror, their eye and the object in each case. Ask them to show and explain how they saw the object (by indicating the path of the light). Can they see round a corner using the mirror? Is the reflection the "same way round" as the object? For older children, ask them to bend the mirror – or use curved mirrrors – and see if that helps them to see more or fewer objects. What do they think is happening to the light path? Is this like any mirror they know in an everyday situation?

Ask children to look at some writing in a mirror. Ask them to describe what it looks like and whether it is easy to read. What seems to have happened to it? Get them to write something so that it looks normal in the mirror. Ask them the difference between mirror writing and normal writing. Do they know of an example of mirror writing in everyday life (such as an ambulance). For older children, ask them to look again at their reflection. Where does it seem to be? Ask them to move away from the mirror and see what happens to their reflection.

Challenge the children to encode and decode messages in mirror writing.

Note: Some children, who find the concept hard to understand, may be helped by using a real periscope. Not all schools will have one but perhaps one could be borrowed or an inexpensive one purchased.

Safety : Safety mirrors should be used. If glass mirrors are used, they must be backed.
See ASE publication *Be Safe!* for information on all aspects of safety in school science.

Numeracy Skills

Children should be able to :

- estimate, draw and measure angles;
- use a protractor;
- calculate angles in a triangle;
- make patterns by rotating shapes.

Literacy Skills

Children should be able to :

- write a poem based on personal experience;
- experiment with words;
- understand the use of an apostrophe for shortening words.

Shadow

Across my bedroom wall
flapping its giant grey wings:
a monster

Across my bedroom lamp
fluttering its small brown wings:
a moth

SHADOW

Across my bedroom wall
flapping its giant grey wings:
a monster

Across my bedroom lamp
fluttering its small brown wings:
a moth

"Shadow"

Discussion Points

- Do you have a bedroom lamp (show an example or picture of one)?
- What have you noticed on your bedroom walls when the lamp is on?
- Have you ever seen a moth in your bedroom? What did it look like and how did it move?
- Why do you think the two descriptions "Across my bedroom wall flapping its giant grey wings" and "Across my bedroom lamp fluttering its small brown wings" are in the same, very short poem?
- Why are the giant grey wings "giant" and "grey", and the small brown wings "small" and "brown"?
- Why is this poem called "Shadow"?
- How do you think a shadow is made?
- How do you think the size of the shadow changes?

Science Background

- Light travels very quickly about 300,000 km per second. Although that sounds fast it still takes light approximately eight minutes to travel from the Sun to the Earth.
- Light travels in straight lines; this is an important idea for understanding shadows, and quite a logical one, because if it didn't it might be able to bend round objects and corners and then we would not have shadows.
- Shadows are formed because light hits an opaque object (opaque is where light cannot go through) and cannot go round or through it, therefore the area on the other side of the object appears to be darker than the surrounding area.
- The shape of a shadow can be changed by moving an opaque object closer to or further away from the light source. In the poem the moth moves closer and further away and gets smaller and larger giving the appearance of a monster moth.
- Transparent objects allow light to pass straight through them, for example, clear glass.
- Translucent materials allow less light through or reflect or bend the light. Materials such as frosted glass are translucent allowing some light through but not giving a clear image. Many plastics are also translucent.

Key Ideas

- Shadows are produced when light is blocked by an opaque object.
- Shadows vary with the relative distance and direction of object and light source.
- Darkness is the absence of light.
- Light can pass through some materials but not others.

Science Skills

Children should be able to :

- identify the different sources of light;
- understand that light travels and takes time to do so;
- explore and test their ideas;
- make predictions;
- collect evidence;
- use evidence to draw conclusions.

Key Activities

The youngest children can use their hands to create a variety of still and moving shadows on the wall using a powerful torch or overhead projector. Draw round them on a large sheet of paper and ask children to compare their hand shapes.

Make some simple shadow puppets and perform a shadow play; how can the characters be made bigger or smaller?

In the playground on a sunny day, ask children to explore their shadows. Ask them: What shape is it? Where is it? Is it possible to run away away from it? Can it be caught? How can you make a joint shadow with a friend?

Extend the above activity and ask the children to explore how the shape and size of shadows relates to the objects creating them and the distance between object and light source.

Discuss with the children what a shadow is and how it is made. Ask the children to draw their ideas.

Guess or predict what object is producing a particular shadow just by looking at the shadow.

Ask them to look at vertical and horizontal shadows and explore how the length of shadows produced by the sun changes during the day.

Design and make a shadow clock or sun dial: establish the best site for it.

Older or more able children could investigate the relationship between the dimensions of a shadow and the distances between object and light source, and object and "screen". This would be an excellent opportunity to practise drawing graphs, though to get satisfactory results the light source must be small and bright. Does this explain anything about how light travels and what is "in" a shadow? Are shadows flat?

Discuss with children the difference between light and dark. Try to bring out the idea that darkness is the absence of light. How could they make the classroom, or a part of it, dark? What does this do to the light? If practicable, get the children to plan and make a dark area.

Safety : Do not look directly at bright lights.
Never look directly at the Sun without wearing protective eyewear. Electric light bulbs get hot and can burn. Be aware of general electrical safety when using mains-powered light sources.
See ASE publication *Be Safe!* for information on all aspects of safety in school science.

Numeracy Skills

Children should be able to :

- measure and compare accurately;
- understand the properties of two-dimensional and three-dimensional shapes;
- understand the concepts of position and direction;
- understand angles, reflective symmetry, proportions and ratio.

Literacy Skills

Children should be able to :

- use figurative language;
- use the structure of a poem to write an extension, for example, an additional verse;
- use a variety of poetic forms, for example, haiku.
- write in the style of the poet.

The Truth

Some people say
nothing's ever simply black and white
nothing's ever simply this or that
they say the truth lies somewhere
in between.

That's OK
but I read how years ago
some people said the Sun goes round
the Earth
and they tried to kill people
who said the Earth went round the Sun.

I've figured out
in the case of this Earth-and Sun thing
the truth does not lie somewhere
in between.

The Truth

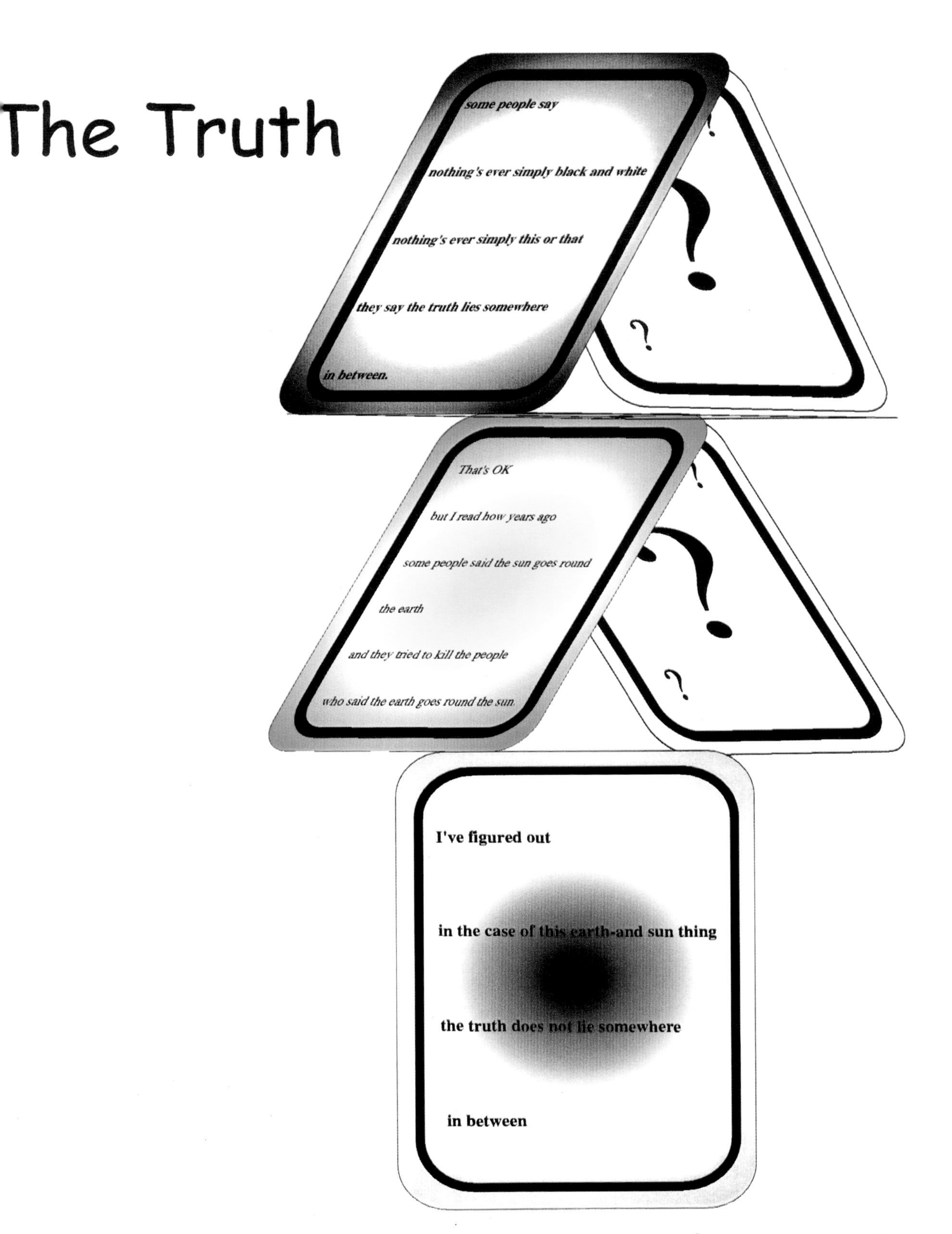

by
Michael Rosen

"The Truth"

Discussion Points

- Have you ever heard someone say that something isn't "black and white"? What were they talking about?
- What do you think the expressions "black and white" and "this or that" mean?
- Can you think of a situation where the truth might "lie somewhere in-between"?
- What is the Earth?
- Do you think the Sun goes round the Earth or the Earth goes round the Sun?
- Do you agree that, in this case, "the truth does not lie somewhere in-between"? Why do you agree or not agree?
- Have you heard of any situation where some people believed one explanation for something harmful while others believed a different explanation?
- How could you prove to someone else that what you believe is true?

Science Background

- The Earth is one of the planets in the Solar System. Each planet orbits (goes round) the Sun. The Earth takes just over a year (about 365 and a quarter days) to make one complete orbit of the Sun.
- The Earth also revolves, or spins, on its own axis. The axis can be visualised as a rod that runs through the Earth from the North Pole to the South Pole. The Earth makes one complete revolution each 24 hours – one day. It is this spinning which causes day and night because each part of the Earth moves through the Sun's light and then into darkness and back again, as the Earth rotates. The Earth's axis is not perpendicular to the plane of its orbit about the Sun. This tilt causes light and temperature variations, which, in turn, cause the seasons of the year.
- Scientific ideas are always changing as new evidence emerges to challenge old theories and superstition. Children should begin to understand that today's scientific knowledge may be different in the future.

Key Ideas

- The Earth is part of the solar system.
- The Earth orbits (goes round) the sun.
- The Earth rotates about its axis.
- The Earth's axis is tilted.

Science Skills

Children should be able to :

- measure and compare accurately;
- record results accurately and in a variety of ways;
- handle and interpret data obtained from secondary sources.

Key Activities

It is impossible to observe the Earth orbiting the Sun directly. What you see is the Sun appearing to move across the sky, at least during the day. Many children therefore believe that the Sun moves round the Earth, as many people did during the first half of the last millennium.

Ask children to gather evidence showing that the Earth orbits the Sun. They could use a variety of secondary sources such as books, video material, computer encyclopedias (which include simulations and NASA pictures) and data-bases to research the accepted view that the Earth orbits the Sun.

This would be a good opportunity for older children to consider the nature of such evidence and how much value can be placed upon it.

Children could report the results of their research to the class and produce a display.

Using an orrery or simple planetarium, you could model the Earth-Sun relationship to small groups. This could be extended or supported by a visit to a science centre or to a museum with an astronomy section.

Design and make simple models showing the Earth orbiting the Sun, and the Earth rotating on its own axis – night and day.

Some universities and science centres offer illustrated talks by arrangement.

Safety : The most obvious hazard is looking at the Sun. Never look directly at the Sun without wearing protective eyewear.
See ASE publication *Be Safe!* for information on all aspects of safety in school science.

Numeracy Skills

Children should be able to :

- understand the properties of two-dimensional and three-dimensional shapes;
- understand the relationship between an object's position and the direction in which it moves;
- understand angles, proportions and ratio.

Literacy Skills

Children should be able to :

- research information about the Sun and Earth;
- use a flow chart to present the information;
- construct a persuasive argument;
- role play the Earth moving around the Sun.

Boogy Woogy Buggy

I glide as I ride
in my boogy woogy buggy
take the corners wide
just see me drive
I'm an easy speedy baby
doing the baby buggy jive

I'm in and out the shop
I'm the one that never stops
I'm the one that feels
the beat of the wheels
all that air
in my hair
I streak down the street
between the feet that I meet

No one can catch
my boogy woogy buggy
no one's got the pace
I rule this place
I'm a baby who knows
I'm a baby who goes, baby, goes.

Boogy woogy buggy

I glide as I ride
in my boogy woogy buggy
take the corners wide
just see me drive
I'm an easy speedy baby
doing the baby buggy jive

I'm in and out the shop
I'm the one that never stops
I'm the one that feels
the beat of the wheels
all that air
in my hair
I streak down the street
between the feet that I meet

No one can catch
my boogy woogy buggy
no one's got the pace
I rule this place
I'm a baby who knows
I'm a baby who goes, baby, goes.

"Boogy Woogy Buggy"

Discussion Points

- What does the child in the poem think is so special about his buggy?
- What can make a buggy hard or easy to push?
- What words are used to describe the different ways the buggy moves?
- What does a buggy have that helps it to move?
- Does anyone have a "Boogy Woogy Buggy" at home? What makes it special?
- Why do you think is it easier to push than to carry?
- Can the buggy be stopped in different ways? What are they?
- What is special about a buggy that makes it easy to change the direction it is facing?

Science Background

- Early experiences with forces in terms of movement help lay the foundations of later, more complex ideas on forces.
- Thinking about the way that they and other objects move helps children to develop the idea that one's own body movement and the forces it exerts affect other things. "I push, the buggy moves, a bigger push makes the buggy move further."
- This develops the idea that the change in the movement of the buggy is related to the direction and the size of the force.
- Buggies are made from materials that are durable and easy to clean, light – so they can be easily carried, and strong – to withstand bumps and scrapes.

Key Ideas

- Observe and describe different ways of moving.
- Explain how familiar objects move.
- Movement can be speeded up, slowed down or stopped.

Science Skills

Children should be able to :

- suggest ideas;
- conduct a fair test;
- communicate what happened during their work;
- make comparisons;
- explain what they have found out;
- cooperate with others.

Key Activities

Look at the wheels on a number of pushchairs and toy cars, and other things that move. Ask children to compare similarities and differences. Which moves the easiest? Which takes the most push to get it going? Once going which gives the smoothest ride? Run the wheels through damp sand and compare tread marks. How do toys or pushchairs move over different surfaces?

Ask the children to classify toys according to the way that they move, for example: wind up, wheels, pull backs etc.

They could explore the pathway of shapes and objects to see how they roll. What path do they follow? What kind of shape rolls the furthest? Ask the children to predict which will or will not roll and which will roll the furthest? Ask children to group their findings and draw a simple table.

Attach different and/or similar shapes to the ends of card tubes (axle) and see which travels furthest? Describe the way the tube moves.

Using cogs of different sizes from construction kits explore how moving one cog can move another. Which direction do they go? Do some go faster than others?

Using wind up toys compare: the way they move; how they are enabled to move. Watch their pathways. Which goes furthest or quickest? Which runs for the longest? Watch the movement of the toys. Is the movement steady in pace? Where is it at its fastest or slowest?

Put a toy car down a slide. Observe its landing. Is it safe? How far does it travel? Can the children think of a way to slow it down in some way? Explore how many ways a car can be slowed down on the slide.

Safety : Care must be taken if children are exploring slides. They should help to decide on safety rules. If putting objects down slopes, good clearance is needed at the bottom.
See ASE publication *Be Safe!* for information on all aspects of safety in school science.

Numeracy Skills

Children should be able to :

- use Venn diagrams to sort toys;
- measure distance using standard and non-standard measures;
- estimate distances;
- compare measurements, for example, how far toys move.

Literacy Skills

Children should be able to :

- identify patterns of rhythm and rhyme;
- judge the effectiveness of the words used;
- experiment with words;
- read poems so they make sense.

Catapults

No one laughed.

For days
we had been winding elastic bands
round our fingers
to make catapults.

For days
we had been folding paper
into different shapes
to make pellets.

For days
we had been pulling back the elastic
to fire off at each other.

This time
no one laughed.

Wendy Sutton opened the window.
Jeff Clarke fired the pellet.
The pellet hit Wendy Sutton.
Wendy Sutton lost the sight in her right eye.

We stopped making elastic band catapults.

Catapults

No one laughed. For days we had

been winding elastic bands round our fingers to make catapults

For days we had been folding paper into different shapes to make

pellets For days we had been pulling back the elastic to fire off at eac

other This time no one laughed Wendy Sutton opened the window

Jeff Clarke fired the pellet The pellet hit Wendy Sutton Wendy Sutto

ost the sight in her right eye

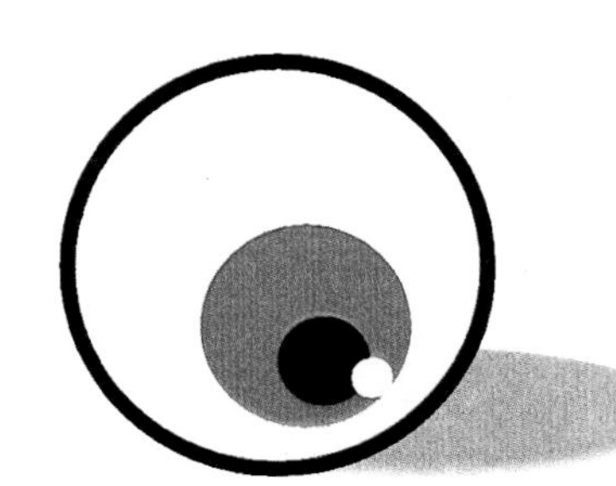

We stopped making elastic band catapults.

"Catapults"

Discussion Points

- How did the children make a catapult?
- How does a catapult work?
- What happened in the poem when the pellet was fired?
- How did Jeff make the pellet fly across the room?
- Where did the pellet get its energy from?
- How can you play with catapults safely?
- How would you make a catapult that made an object travel a long way?
- What is special about the materials used to make a catapult?
- Why should you not fire catapults at people, animals or delicate objects?

Science Background

- An object will change shape when a force is exerted on it.
- An elastic object will go back to its original shape when the forces that have changed its shape are released; this is elastic potential energy.
- Metal springs are made in such a way so as to behave as though they are elastic.
- Both elastic bands and springs have an elastic limit.
- When an elastic band in a catapult is pulled back and held, the forces are in balance. The missile is released from a position of rest. As the rubber band is released it finds its original shape. The missile slows down as other forces, for example, gravity and air resistance, act upon it.

Key Ideas

- Springs can be found in many everyday objects and their ability to compress and stretch makes them useful.
- The stretch or compression exerted on an elastic band or spring can be measured.
- When a force is exerted on a spring or elastic band it will want to return to its original shape; this change can be used to exert a force on another object.

Science Skills

Children should be able to :

- plan and carry out a fair test;
- make predictions based on knowledge;
- work with others.

Key Activities

Ask the children to pull elastic bands of different sizes with their hands. How far do they stretch? In which direction are the forces involved? Can they feel the force coming back from the elastic band? Does the size, length or thickness of the elastic band make a difference?

Tie two wooden blocks together, using the strongest feeling elastic bands. Pull the blocks until the elastic bands are fully stretched and then let go. What happens to the blocks? Can the children see any pattern in the distance stretched and the movement of the blocks?

Look carefully at objects that use a spring (ball point pen, mattress, jack-in-a-box, stapler). Identify when the spring is in compression or tension (stretch). Identify the forces that are acting on the spring and the forces created by the spring. Draw pictures with force arrows.

Give the children a variety of springs of different sizes or with different coils and ask them to squeeze the springs between finger and thumb. Can they feel the force that is needed to compress the springs and the force exerted back through the fingers from the spring? Do all the springs have the same amount of force? What seems to make the difference to its ease of compression and force? Ask the children to think of places where these different springs would be useful.

Explore compression by using a large (but not too strong) spring secured to a base-board. To prevent the spring falling over, place it inside a transparent container, such as a plastic lemonade bottle with the top cut off. Add masses on top of the spring one at a time (approx. 100g but the amount will depend on the spring) and measure how much the spring compresses. Ask the children to predict the next measurement. Draw a line graph of the results and estimate readings for interim masses.

Using the method below, explore how many masses are needed to stretch different springs by a given length.

Add 50g masses (experiment beforehand to find a suitable mass) to a given spring and record the stretch on a line graph.

Repeat the experiment, only this time hanging 100g masses on a spring or elastic band. Make a Newton-meter by attaching the spring to a board and marking the graduations on to the board. The force exerted by 100g is approximately one Newton.

Explore wind-up and pop-up toys. Which go fastest, travel the furthest, or jump the highest? Will they go up a slope or pull another object? Explore the relationship between the number of windings and the distance travelled by a wind up toy.

Safety : Elastic bands and springs need to be handled with great care. Wear safety goggles.
See ASE publication *Be Safe!* for information on all aspects of safety in school science.

Numeracy Skills

Children should be able to :

- make relevant observations and accurate measurements;
- understand why observations and measurements may need to be repeated;
- record results, make deductions and arrive at conclusions based on the results;
- present findings on graphs and tables;
- read and interpret graphs.

Literacy Skills

Children should be able to :

- create a play using the story in the poem;
- produce a safety poem for the classroom;
- write a set of instructions for an investigation.

Footsteps

Trod, tread, trudge, traipse and tramp
stagger, step, scamper, stump and stamp;
Hike, skip, jog, march and amble;
stroll, strut, shuffle, stride and ramble;
Saunter, lope, dawdle, plod and toddle;
potter, run, wander, walk and waddle.

There are many ways of
moving along on foot
and just as many ways the matter can be put.

Footsteps

Trod, tread, trudge, traipse and tramp stagger, step, scamper, stump and stamp; Hike, skip, jog, march and amble; stroll, strut, shuffle, stride and ramble; Saunter, lope, dawdle, plod and toddle; potter, run, wander, walk and waddle. There are many ways of moving along on foot and just as many ways the matter can be put.

"Footsteps"

Discussion Points

- What ways of moving does the poem describe?
- Which words tell us about moving slowly or quickly?
- Which words are words we sometimes use to describe animals moving? Which words for which animals?
- What other words can you think of to describe the way humans move?
- In what ways do plants, animals or insects move?
- Look at some non-living objects – can you think of words to describe the way that these move?
- How do non-living objects move?
- Where do animals, including humans, get their energy to move? How is this different to objects such as toy cars and elastic bands?

Science Background

- Teachers often talk about healthy living and eating and discuss diet, exercise and drugs with children but rarely mention feet. Yet many adult ailments have their origin in childhood where neglect of the feet can lead to poor posture. This may result in problems in other parts of the body such as the legs and the back.
- Young feet are soft. Inadequate foot care and poorly fitting shoes can lead to pressure on the foot and deformities in the first years of life. This is particularly important since children's feet can grow to half their adult foot size during the first year. Later in life adults may suffer problems because of lack of foot care in the early years and also from shoes that have been purchased because they are cheap, or to follow the latest fashion trend; they are not always chosen for the best fit and comfort.
- In the human foot there are at least 26 bones, plus ligaments, muscles, blood vessels and nerves. It is a complex part of the body and important because the feet carry the entire weight of the body for the lifetime of most people.
- The instep has five elongated bones that join the ankles. The ends of the bones form the ball of the foot. There are also ligaments that form the arches of the feet and are the stable spring base for human feet.
- Not all animals have the same feet for movement. Some animals have adapted the basic five toes. The horse, for example, has only one toe which has a hoof; the hoof is the equivalent of a human toenail.

Key Ideas

- Know that animals need to move to stay alive, for example, to catch food.
- Know about the skeleton and bones in the legs and feet.
- Know the role of muscles in movement.
- Know about using energy to move.

Science Skills

Children should be able to :

- suggest ideas;
- recognise a fair test;
- explore using senses;
- communicate what happened during their work;
- make comparisons;
- try to explain what they have found out;
- cooperate with others.

Key Activities

Let children try travelling in all the different ways mentioned in the poem. What parts of their feet or body are in contact with the floor? Can they speed up or slow down all of the actions? Why? Why not?

Cover the soles of the children's feet with talcum powder and let the children walk across a strip of black paper in talc covered feet.

Do they all walk in the same way?
What part of the foot is used most?
Who has the biggest stride?
Who puts the smallest area of their feet down?
How could the children find out?
Compare the footprints of a run to a jump.

Younger children can listen to a tape recording of people walking. See how many footstep sounds they can identify and if they can distinguish footsteps walking on different surfaces. Use the desciptions in the poem as a prompt.

Consider other animals. Look at pictures of them. Does what they look like give any clues to the way they move?

Using real animals (pets or a video) look at the way they move. Do they have legs? Do they walk on two or four legs? Are their forelimbs used for other things as well as walking?

Safety : Be aware of asthmatics when using talcum powder.
See ASE publication *Be Safe!* for information on all aspects of safety in school science.

Numeracy Skills

Children should be able to :

- ◆ measure and calculate the area of different shapes, for example, footprints.
- ◆ find different ways to calculate the area of a shape;
- ◆ find the perimeter of a shape, for example, a footprint.

Literacy Skills

Children should be able to :

- ◆ understand where to use commas;
- ◆ use rhythm and rhyme;
- ◆ use descriptive language;
- ◆ understand alliterative use of language;
- ◆ extend vocabulary by using a thesaurus;
- ◆ identify adjectives in a piece of writing.

Fridge magnets

After our holiday in the States
we came back with fridge magnets:
a tiny bubble gum dispenser that rattles
a juke box that plays Rock around the Clock
an English telephone box that rings
an American payphone that rings
a kitchen blender that whirrs
a knickerbocker glory
and
a poem

When my sister comes in
she rattles the tiny bubble gum dispenser
she plays Rock around the Clock
she rings the English telephone box
she rings the American payphone
she whirrs the blender
she licks the knickerbocker glory
and she reads
the poem

She takes a drink from the fridge,
slams the door and
the tiny bubble gum dispenser that rattles
the juke box that plays Rock around the Clock
the English telephone box that rings
the American payphone that rings
the kitchen blender that whirrs
and the knickerbocker glory
drop off the fridge door.

The poem stays there.

Fridge Magnets

After our holiday in the States
we came back with fridge magnets:
a tiny bubble gum dispenser that rattles
a juke box that plays Rock around the Clock
an English telephone box that rings
an American payphone that rings
a kitchen blender that whirrs
a knickerbocker glory
and
a poem

When my sister comes in
she rattles the tiny bubble gum dispenser
she plays Rock around the Clock
she rings the English telephone box
she rings the American payphone
she whirrs the blender
she licks the knickerbocker glory
and she reads
the poem.

She takes a drink from the fridge
slams the door and
the tiny bubble gum dispenser that rattles
the juke box that plays Rock around the Clock
the English telephone box that rings
the American payphone that rings
the kitchen blender that whirrs
and the knickerbocker glory
drop off the fridge door.

The poem stays there.

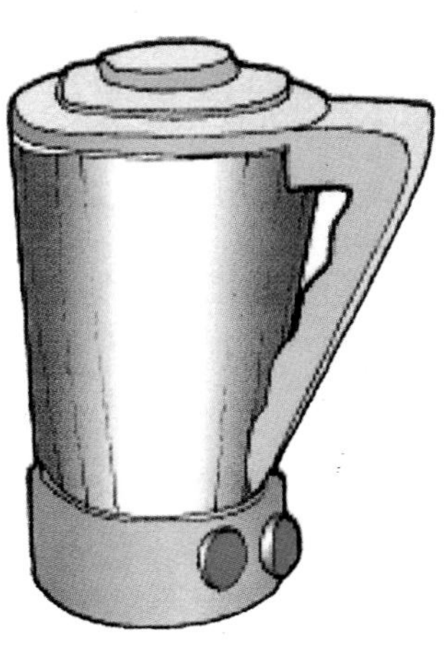

"Fridge Magnets"

Discussion Points

- What is a fridge magnet?
- Do you have any special ones – like those in the poem?
- What do you think makes them stick to the fridge door?
- What else will they stick to?
- Why do you think nearly all of them fell off when the sister slammed the fridge door?
- Why do you think the poem didn't fall off?
- Which materials will the magnet attract?
- How can you find out which is the strongest magnet?
- How could you make your own magnet?

Science Background

- A magnet has a magnetic field all round it which gets weaker with increasing distance from the magnet. On a bar magnet the magnetic field seems to be strongest around each end, or "pole", of the bar.
- When something magnetic, such as an object made of steel or iron, comes within a magnet's field, there is an interaction resulting in an attractive force between them.
- If the fields of two magnets "overlap", the interaction results in either an attractive or a repulsive force depending on the relative alignment of the two magnets. We say, "Like poles repel, unlike poles attract".
- Magnets have many uses in everyday life, for example: catches on cupboards, around fridge doors, handbag fasteners, strips for holding kitchen knives.

Key Ideas

- Some materials are magnetic and others are not.
- There is an attracting force between magnets and magnetic materials.
- There are attracting and repelling forces between magnets.

Science Skills

Children should be able to :

- understand and follow instructions;
- consult books and databases;
- organise information in a orderly way;
- record results for reports or display;
- work with others.

Key Activities

Ask a small group of children to test everything in the classroom with a magnet and find out what happens. Perhaps they could use a fridge magnet and/or a small bar magnet.

Ask the children to make two sets of drawings, one of "magnetic" and the other of "not magnetic" things.

Discuss with them which things would be good for "sticking" a fridge magnet to?

Take the children through the following activities and ask them for their reactions.

Bring a magnet very slowly near to something magnetic. What do you feel? Can you hold the magnet very close to the magnetic material without letting them touch? Why not?

Bring one end of a bar magnet very slowly near to one end of another bar magnet. What do you feel?

Turn one of the magnets the other way round and do it again. What do you feel this time?

When the magnets are attracting each other, can you hold them very close together without letting them touch? When they are repelling, how can you make them touch?

Float a bar magnet on some expanded polystyrene on water in a sink or trough. Using another magnet and without touching the magnet that is floating on the polystyrene, can you make the floating one move towards you and away from you, as you choose? How did you do it? Can you explain what happened?

Challenge the children to design and make simple magnetic toys or games such as magnetic fishing or magnetic football.

Safety : See ASE publication *Be Safe!* for information on all aspects of safety in school science.

Numeracy Skills

Children should be able to :

- use a Venn diagram to sort magnetic and non-magnetic items;
- understand intersecting circles;
- draw conclusions using the information from a Venn diagram.

Literacy Skills

Children should be able to :

- discuss the differences between rhyming and non-rhyming poems;
- use alliteration, for example, "magic magnets";
- express views about a poem.

Granma's Hands

My Granma's 72
She says she's getting old.
My Granma's 72
She says her hands are cold.

My Granma stands at the bus-stop
in all kinds of weather
My Granma stands at the bus-stop
rubbing her hands together.

'I always rub me hands,' she says
'when I'm running from pillar to post
'I always rub me hands,' she says
'ohh, they're warm as toast.'

Related poems:

Engine Oil
Lubricate the Joints

Granma's Hands

My Granma's 72
She says she's getting old.
My Granma's 72
She says her hands are cold.

My Granma stands at the bus-stop
n all kinds of weather
My Granma stands at the bus-stop
ubbing her hands together.

'I always rub me hands,' she says
'when I'm running from pillar to post,
'I always rub me hands,' she says
'ohh, they're warm as toast.'

"Granma's Hands"

Discussion Points

- How many of you have waited at bus stops in the cold?
- What kind of things do you do to keep warm?
- What does grandma do to keep warm?
- First, rub your hands together slowly, then rub them together quickly.
- What happens? What can you feel? What can you see?
- What do you think makes your hands feel hot?
- What can you do to make them even warmer?
- Can you think of any other examples where two surfaces rubbing together creates heat, such as car tyres at the end of a journey?
- What different ways do people use to keep warm in winter?

Science Background

- Friction is the force created by two surfaces moving against each other.
- When two objects are rubbed against each other, heat is produced as a by-product. This is because energy is being used to overcome the friction and some of this energy is released as heat.
- Friction is a force that opposes movement but can either ease or hamper the movement of objects and materials.
- Without friction everyday movements become impossible, for example: friction occurs when writing with a pen or pencil on a piece of paper; a rolling ball slows down or stops because of friction.
- The surface does not have to be a solid. Friction also occurs through movement in both liquids and gases.
- All solids have tiny irregularities that catch and hinder movement. If the surface is lubricated these irregularities are easier to resist and friction is reduced.

Key Ideas

- There is a force between two moving surfaces in contact called friction. The force may or may not be useful.
- Different surfaces cause different amounts of friction to a moving object.
- Lubricants reduce the effects of friction and allow surfaces to move more freely.

Science Skills

Children should be able to :

- turn suggestions into ideas;
- understand the use of control experiments;
- investigate, predict and conduct a fair test;
- choose and use apparatus and equipment accurately;
- draw conclusions;
- work with others.

Key Activities

Observe the effects of friction on balls, bean bags, or quoits by watching their movement across different surfaces. Can children identify the different speeds the objects travel? Do they roll or skid? Brainstorm examples from everyday experience. Perhaps focus on a fairground or playground where friction is either helpful or can cause problems.

Conduct a rubbing test. Rub a block of wood covered with material against different surfaces to test which surfaces create holes in a given material. This is a simple measure of the friction created by the two surfaces. Alternatively, rub a variety of materials on the same surface to investigate which is the safest material to wear in a playground to protect against friction burns.

Using a slope, slide, marble run, or guttering, explore how the speed of an object can be changed as it travels down a helter skelter by changing the helter skelter's surface. The surface can be made: wet, dry, oily, sandy, smooth or rough. The children should be encouraged to keep the object and the slope the same.

Using different lubricants, for example: vaseline, washing-up liquid, polish, baby oil ,vegetable oil, or water (plastic gloves should be used when applying lubricants) – use a Newton-meter or stretch an elastic band to measure the force needed to move an object across a given surface. To obtain different readings make sure the surface is not too smooth. Compare the results against a non-lubricated surface as a control.

Explore other methods of reducing friction including rollers, marbles or wheels.

Safety : Be aware of the hazards when handling lubricants. Wear safety goggles.
See ASE publication *Be Safe!* for information on all aspects of safety in school science.

Numeracy Skills

Children should be able to :

- measure using a Newton-meter;
- use a stopwatch accurately;
- take repeated readings;
- calculate mean, mode and median
- record results on charts or tables.

Literacy Skills

Children should be able to :

- understand how to use both rhyming and non-rhyming words;
- understand the use of direct and reported speech;
- write a verse in the style of the poem.

High Rise Ships

At night
in the harbour
the ships tower above us

like high rise blocks of flats
I think
they're great buildings
with kitchens and bedrooms
and a thousand stairs

Inside
there are car parks full of cars
and the windows are lit up
across and down
and right up to the top
just like office blocks
and city buildings

All that steel and glass

Humming in the middle
you can hear the mass of engines
turning and turning

And if you wait
there comes a moment
when these buildings move

They shudder
and nudge out
into the harbour

And you see clearly now
across the water
that this great building
dares to float

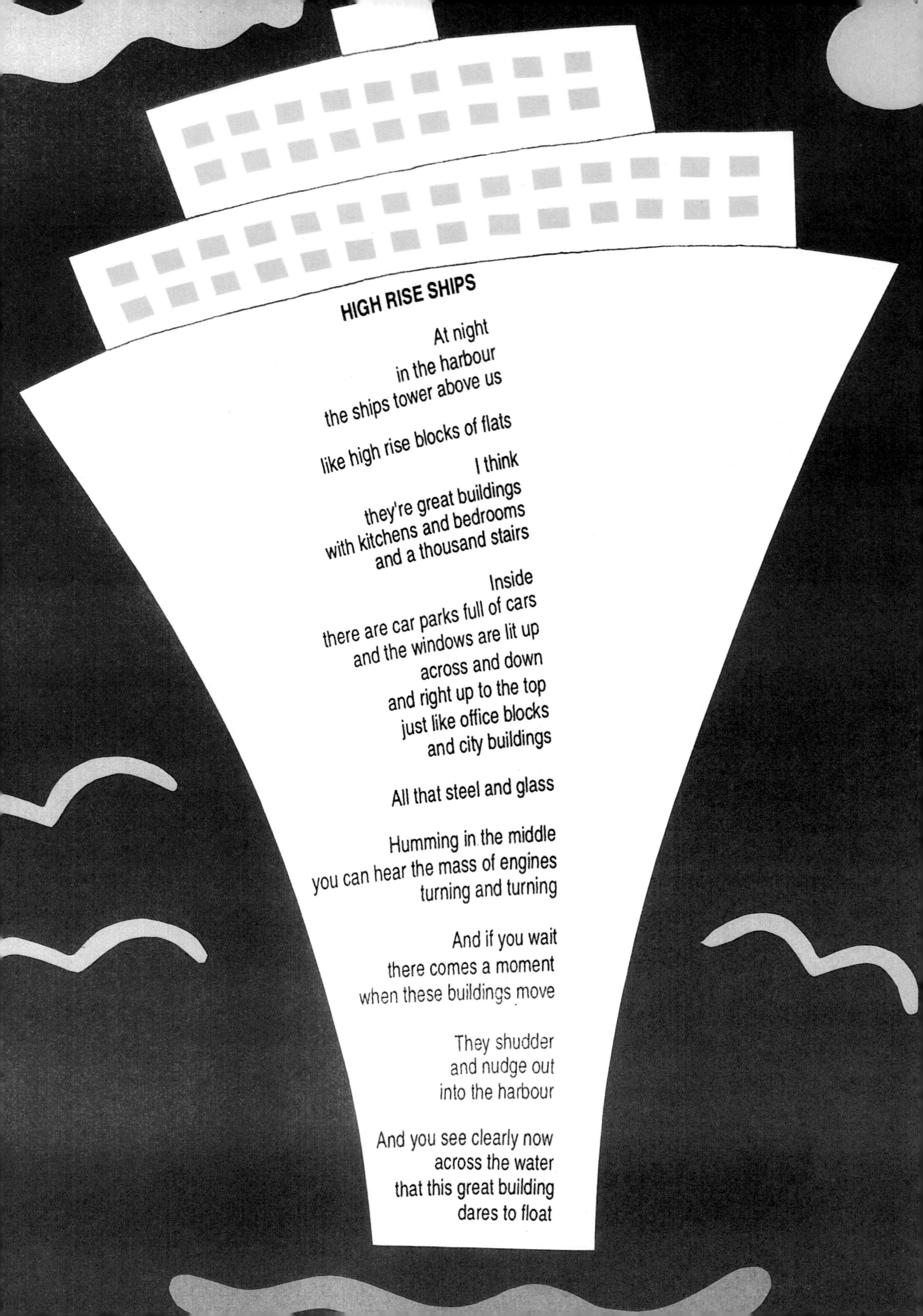
HIGH RISE SHIPS
At night
in the harbour
the ships tower above us
like high rise blocks of flats
I think
they're great buildings
with kitchens and bedrooms
and a thousand stairs
Inside
there are car parks full of cars
and the windows are lit up
across and down
and right up to the top
just like office blocks
and city buildings
All that steel and glass
Humming in the middle
you can hear the mass of engines
turning and turning
And if you wait
there comes a moment
when these buildings move
They shudder
and nudge out
into the harbour
And you see clearly now
across the water
that this great building
dares to float

"High Rise Ships"

Discussion Points

- What do you think it is like on such a big ship?
- How does the person who is looking at the harbour describe the ship? Why?
- What materials are the ships in the poem made from?
- What size do you think the ship is?
- Why do you think he or she says that "this great building dares to float?"
- What do you think he or she thinks should happen to the ship and why?
- What objects can you think of that float?
- What kind of big things have you seen floating?
- Why do you think they were able to float?
- What do you think makes objects sink?

Science Background

- All objects experience a force due to gravity pulling them towards the centre of the Earth.
- This force is proportional to the mass of the object and is known as weight. Objects partly or wholly in a liquid or fluid, usually water, experience a force opposing their weight. This force is proportional to the mass of the water or fluid displaced or pushed out of the way by the object. This is known as upthrust.
- If the upthrust is equal to the object's weight it will float. If it is less than the weight of the object it will sink.
- If an object has a large surface area in contact with the water, the surface tension or "skin" effect holding the object on to the surface of the water is significant.

Key Ideas

- Objects float in water when the upthrust is equal to the gravitational pull.
- Floating and sinking depends on a number of factors including shape, material and mass.
- Objects can float or sink in liquids other than water.

Science Skills

Children should :

- be able to make accurate observations and measurements;
- know what questions to ask and why;
- be able to draw conclusions from their investigations and apply these conclusions to everyday life;
- be able to work with others.

Key Activities

Use children's suggestions for why objects float or sink arising from the High Rise Ships poem and provide resources for them to explore their ideas.

Ensure you include objects that question any misconceptions arising from the poem, such as heavy or large things that float and small, light things that sink. Explore plastic bottles with different contents. For example, put plasticine of the same weight in containers of various shapes. Do the children realise the weight is constant?

Repeat the test in different volumes or depths of water; children often think that a bigger volume of water aids flotation.

Test buoyancy aids, such as inflatable arm-bands or rings. Use a weight to represent a child whose "weight" in water is close to zero. Tie the weight to the buoyancy aid using string. How inflated do the arm-bands or rings have to be to enable the weight to float? What is it that changes to enable the weight to float?

Compare different sized arm-bands. Does the weight they are able to support vary? Why do arm-bands often have a weight or age written on them?

What implications do their findings have on water safety? Results could be recorded in the form of a letter of advice to new parents or owners of swimming pools. Include in the letter details of the tests carried out and the findings.

Measure objects that sink using a Newton-meter, by attaching the object to string and suspending it, firstly in the air, and secondly in water. The reading is less in water as the upthrust of the water acts on the object. Record the findings on a chart. Encourage three readings of each measure to ensure accuracy. Floating objects have a reading of zero Newtons. Can children begin to predict readings in the water? Ask them to explain their readings. Can they draw diagrams with force arrows to show what happens?

Using the idea of a floating bridge, put different weights on a polystyrene "floating bridge" and explore at what point it sinks. Does it matter how the weight is distributed? Children can measure the amount of polystyrene under and above the water with each weight and make predictions for its sinking point. Ask them to apply this idea to limits on the number of cars that could be carried on a ferry.

Ask the children to investigate boats of different heights. Are some more stable than others? How could they make a tall ship stable? Challenge them to design and make a boat that can carry a given load across a given distance.

Safety : Recognise, assess and take action to control risks in everyday contexts. Even small amounts of water can represent a safety risk.
See ASE publication *Be Safe!* for information on all aspects of safety in school science.

Numeracy Skills

Children should be able to :

- use appropriate measuring equipment, for example, a Newton-meter;
- read different scales;
- record measurement using appropriate units.

Literacy Skills

Children should be able to :

- recognise similes;
- understand the use of prepositions;
- use formal and informal language;
- use descriptive language.

Who Stuck the Sofa?

It wasn't my fault
the sofa got stuck in the door.

You see, my Dad said
it had come in through the door
so it was obvious
it had to be able to get out the same
way.

So he was up one end
and my sister and me were up the
other
and we headed for the door.

It looked like it would go through
on its side
kind of leaning over.
He pulled.
We pushed.
But it wouldn't go.

We pushed.
Then it looked like it would go through
standing up
kind of twisted round.
We pulled
He pushed.
But it wouldn't go.

Then it looked like it would go through
upside down
kind of legs first.
He pulled.
We pushed.

And it really was looking good
it was halfway through that doorway
when my Dad said,
'what's it look like your end?'

And it looked really good.
Really it did.

so I said,
'OK'.

Why did I say
'OK'?

If I could have my time over again
the one thing
I would NOT say, is
'OK'.
But I did.
so he pulled
And we pushed
And - KER-unfffff.
It stuck.
In the doorway
The doorway of the front door.

Where everyone comes in
and everyone goes out.

That is
until the sofa got stuck there
because I said,
'OK'.

It's been there for two days now.

The postman delivers letters under it.
Next door's kids climb over it.
The cat sleeps on it.

And Dad is trying to make up his
mind
what to do about it.

It wasn't my fault.
Really it wasn't.
All I said was
'OK'.

Who Stuck the Sofa?

It wasn't my fault
the sofa got stuck in the door.

You see, my Dad said
it had come in through the door
so it was obvious
it had to be able to get out the same way.

So he was up one end
and my sister and me were up the other
and we headed for the door.

It looked like it would go through
on its side
kind of leaning over.
He pulled.
We pushed.
But it wouldn't go. We pushed

Then it looked like it would go through
standing up
kind of twisted round.
We pulled.
He pushed.
But it woudn't go.

Then it looked like it would go through
upside down
kind of legs first.
He pulled.
We pushed.

And it really was looking good
it was halfway through that doorway
when my dad said,
'what's it look like your end?'

And it looked good.
Really it did.
so I said,
'OK'.

Why did I say
'OK'?

If I could have my time over again
the one thing
I would NOT say, is
'OK'.
But I did.

so he pulled
And we pushed.
And - KER-unfffff.
It stuck.
In the doorway
The doorway of the front door.

Where everyone comes in
and everyone goes out.

That is
until the sofa got stuck there
because I said,
'OK'.

It's been there for two days now.

The postman delivers letters under it.
Next door's kids climb over it.
The cat sleeps on it.

And Dad is trying to make up his mind
what to do about it.

It wasn't my fault.
Really it wasn'n.
All I said was
'OK'.

"Who Stuck the Sofa?"

Discussion Points

- Have you ever tried to move a sofa?
- Why were Dad and the children pushing and pulling the sofa?
- Who was pushing and who was pulling?
- In what direction(s) were they pushing and pulling – the same direction or different directions? Why?
- What happened to the sofa?
- Why did it get stuck?
- What was holding it there?
- What forces were in action?
- How would you get the sofa through the door?

Science Background

- The effects of forces are not abstract. Force or forces are acting when the movement of an object is changed. A force acting on an object can make it start moving, stop moving, speed up, slow down or change direction.
- Force may also change the shape of an object, as in a stretched elastic band.
- Pushes and pulls are good examples of forces acting which children can recognise and feel easily. Also the direction in which they act is usually obvious.
- If an object is stationary or moving at a constant speed, the forces acting on it are balanced (usually equal in size and opposite in direction), for example, when an object is floating.
- The commonest examples of balanced forces relate to stationary objects, although it is not always obvious to children which forces are acting. You can help by getting the children to understand that weight is a force. Weight is how we describe and measure the action of gravity on the mass of an object, pulling it towards the centre of the Earth and therefore the ground.
- When an object is stationery the forces acting on it are balanced. In the case of the sofa stuck in the door the forces are balanced and would need a big push or pull to move the sofa.

Key Ideas

- The force needed to move an object depends on the weight of the object.
- There is always resistance to the force being used.
- If resistance and force are equal the object will not move.

Science Skills

Children should be able to :

- explore forces;
- use a Newton-meter to measure a pull force;
- record data;
- use data to make generalisations.

Key Activities

Ask children individually to push and pull on fixed equipment such as wall-bars. What do they feel and what happens? Does anything move?

Then get them to work in pairs, pushing and pulling each other, but not too vigorously. What do they feel and what happens? They should be able to feel the force and notice that sometimes they move and sometimes they don't. This will depend on whether the forces balance or not.

Ask a group to move a piece of equipment, such as a bench, to a specific spot by pushing and pulling it. What arrangement works best? What directions do they have to push and pull to move the bench to the right place? Can they push and pull so that the bench does not move?

Using a model "sofa", children can explore, qualitatively and quantitatively, the forces – pushes and pulls – needed to move it in a specific direction.

Measurements could be taken using a Newton-meter. These conventional units could be used, and results recorded in a table.

Make simple models of a sofa from different materials such as plasticine, clay, card, wood and lego. Explore what happens when, say, a block of wood (representing the person in the poem) is pushed into each one with the same force.

Note: Physical education sessions would be ideal for this activity because the children have space to move around safely and can be supervised carefully.

Safety : No specific hazards relating to the science activities or resources, but the usual safety precautions for children doing physical education and other outdoor activities
See ASE publication *Be Safe!* for information on all aspects of safety in school science.

Numeracy Skills

Children should be able to :

- understand the concepts of position and direction;
- understand angles, proportions and ratio;
- choose a forcemeter with a suitable scale and use it properly;
- record readings accurately and create a graph of the results.

Literacy Skills

Children should be able to :

- recognise and use verbs;
- understand the use of tenses;
- understand the use of direct and indirect speech;
- use synonyms;
- use a thesaurus.

Rubber Dubber

rubber dubber
flouncer bouncer
up the wall
and in and outer
under over
bouncing backer
mustn't stopper
in betweener
do a clapper
in betweened
do a spinner
faster faster
to and fro-er
rubber dubber
flouncer bouncer

BUT

then oh bother!
Butter finger
dropped the ball
and pitter patter
patter pitter
rubber ball
ran right away.

Rubber Dubber

rubber dubber
flouncer bouncer
up the wall
and in and outer
under over
bouncing backer
mustn't dropper
mustn't stopper
in betweener
do a clapper
in betweener
do a spinner
faster faster
to and fro-er
rubber dubber
flouncer bouncer

BUT

then oh bother!
Butter finger
dropped the ball
and pitter patter
patter pitter
rubber ball
ran right away.

"Rubber Dubber"

Discussion Points

- What do you think Rubber Dubber is? What does it look like? Why is it "flouncer bouncer"?
- What do you think it is made from? What is special about rubber as a material? How can you find out where rubber comes from?
- Make a collection of balls and discuss their similarities and differences.
- Do all balls need to be bouncy? What kind do not?
- Are balls made of solid material or do they sometimes have something else inside?
- What do they have inside them? Why?
- What is an "in and outer" and an "under over"? How do you do "a clapper" and "a spinner"?
- What do you think "butter finger" means?
- After the ball was dropped, how did it roll right away? What made it roll?
- When you drop a rubber ball on the floor what happens to its shape?
- What happens if you try to squash a rubber ball between your fingers?
- Where does the force come from that makes the ball bounce back up?
- What kind of materials do you think would make a good Rubber Dubber?

Science Background

- Elastic means a material that when stretched will return to its original shape.
- Rubber is an example of a material that when a force is applied and energy has been used to stretch it, it will absorb and store the energy. When the material is released, the stored energy can be used to propel an object and return the material to its original shape.
- Different kinds of rubber have different properties and therefore different uses. For example, silicone rubbers remain stable over a range of temperatures and therefore are used for oven door seals, tyres in aircraft and even heart valves.
- However, when a force (for example, hitting with a hammer) is applied to materials with different properties other things happen. For example glass, which is brittle and a poor absorber of energy, can crack, whereas steel can absorb the energy and keep its shape.
- When an object such as a golf ball or a football hits another surface a force is generated. This force pushes particles in the ball together making them closer than normal. This generates a repulsive force between the particles which causes them to move back to their original position. We see this as the bounce. If the surface that the ball contacts is soft, such as sand, the particles of sand are also pushed closer together, leaving a dent in the surface.
- In golf balls it is the particles in the material that react; in footballs it is both the material and also the air inside the ball. Since different balls are made of a range of materials and some have more air in them than others, different balls bounce in different ways.

Key Ideas

- Materials are chosen, and may be compared, on the basis of their properties for specific uses.
- When a bouncing ball hits another surface, outside forces push the particles together closer than they would normally be. This generates a repulsion force between the particles. This force causes them to try to move back to their original position.

Science Skills

Children should be able to :

- follow instructions, draw and label diagrams;
- plan and carry out a fair test;
- choose the appropriate apparatus and use it correctly;
- understand why repeat readings need to be taken;
- make predictions based on knowledge;
- work with others.

Key Activities

In these activities children could use ICT spreadsheets to record results. Excel, for example, can be used to average results or plot graphs.

Ask children to investigate which is the bounciest ball. They will need to consider how to do this and how to make it fair. Who will drop it? What will be measured? Will the bounces be counted? Will the test only include balls that are meant to be bouncy?

Children can investigate the factors that affect the bounciness of a selection of rubber balls, including the height from which they are dropped.

They should also feel how much force is needed to squash each ball and how much is "given back" when the ball is allowed to return to its original shape. They can link this to what happens when the ball hits the ground and rebounds.

They should note that each bounce is less than the one before, and so the rebound force must be less. Energy is transferred on impact, particularly to heat – which can be felt when the ball is squeezed repeatedly and allowed to return to its original shape. The general principle can be extended to other elastic and flexible objects.

In these kinds of activities children should understand the principle of repeat readings. Bouncing balls is a good context for teaching this because it is difficult to be exact about how high a ball has bounced. Since we cannot be confident that we have exact readings the appropriate thing to do is to repeat the readings a number of times (usually three to five) and calculate the mean, mode or median.

Challenge children to design and make a device to measure the height to which a ball bounces.

Safety : Bouncing rubber balls requires careful supervision. Safety goggles should be worn. See ASE publication *Be Safe!* for information on all aspects of safety in school science.

Numeracy Skills

Children should be able to :

- make relevant observations and accurate measurements;
- repeat measurements;
- calculate mean, mode and median;
- present findings using tables and graphs.

Literacy Skills

Children should be able to :

- understand the changes in word endings, for example, -er;
- write additional material for the poem;
- prepare a poem for performance;
- use expressions, tone etc.

Scoop a Gloop

Scoop a gloop
of slimy clay
squeeze it, knead it
pummel it, stretch it
roly poly, roll it
into long, thin
sausages.

Bend them, coil them
one on top
of one another
up and up
and round and round
to make
a pot.

It's still soft
and leans a bit
but wait -
and wait -
it slowly hardens
sits dry and dusty
crisp as a biscuit.

Don't tap
or drop
it'll crack
or crumble.

Take it gently
to the kiln
and under fire
of fantastic heat
it strengthens
toughens
enough
to let you
use your spoon
or run your
thumbnail
up and down
your clever coils.

Related poems:

Hot Cat
Some Thoughts About Eggs

Scoop a gloop

Scoop a gloop
of slimy clay
squeeze it, knead it
pummel it, stretch it
roly poly, roll it
into long, thin
sausages.

Bend them, coil them
one on top
of one another
up and up
and round and round
to make
a pot.

It's still soft
and leans a bit
but wait -
and wait -
it slowly hardens
sits dry and dusty
crisp as a biscuit.

Don't tap
or drop
it'll crack
or crumble.

Take it gently
to the kiln
and under fire
of fantastic heat
it strengthens
toughens
enough
to let you
use your spoon
or run your
thumbnail
up and down
your clever coils.

"Scoop a Gloop"

Discussion Points

- What does clay look like when we take it out of the packet?
- What does it feel like?
- What did the poem say it was like?
- What colour is it?
- What do you think it smells like?
- What kind of things is clay good for making?
- How could we make the clay change (permanently) and keep it that way?
- What other materials can you shape with your hands?
- How can you change bread dough and pastry and make them stay changed?
- How can we change other materials by warming or heating them? (not burning)
- Do all these materials remain changed?
- How can they be changed back again?

Science Background

- When materials such as clay, dough, or plasticine are modelled. If the shape is changed and the new shape remains, the material is said to be "plastic". When a force is applied to a material and, afterwards, it returns to its original shape, as with an elastic band, then the material is said to be "elastic". Many common, flexible materials show a combination of the two properties.
- Heating and cooling materials can be associated with a change of state. Heating a solid transfers energy and the molecules move apart creating the liquid state. Further heating adds more energy and movement causing the liquid to change to a gas.
- Generally, heating a material causes an increase in volume and an increase in energy. Cooling causes a decrease in volume and a loss of energy.
- Different materials change state at different temperatures. Water changes to a gas, steam, at 100ºC and to a solid, ice, at 0ºC. Some metals require very high temperatures to change from the solid to liquid state.
- Sometimes heating a material changes it chemically and irreversibly as in cooking an egg or baking a cake. It can also result in burning if the temperature gets very hot. This change sometimes confuses young children as the heat in cooking can change liquid to solid as seen in an egg or wet cake mixture. Albumin or egg white is mostly water but about 9% protein. When heated the protein structure changes permanently; it becomes tighter; it coagulates.
- Water is unusual in that its volume decreases as it cools but as it freezes its volume increases. Also, unlike other materials such as chocolate where the change is gradual, there is no intermediate state between solid and liquid.

Key Ideas

- Use the senses to explore similarities and differences between materials. Describe tactile properties.
- Some objects can be changed in shape by squashing, bending, twisting and stretching.
- Some materials change when heated and cannot change back. The process is irreversible.
- Some materials change when heated and can change back. The process is reversible.
- Materials change state at different temperatures.

Science Skills

Children should be able to :

- use their senses independently;
- describe things accurately;
- communicate with others;
- report findings to the whole class.

Key Activities

Prepare a "feely" bag with different objects inside that have a variety of tactile properties. Encourage the children to feel inside without looking and describe the object. Then pass the bag on. Let a child take out one object and describe what they see, pass the bag on. They could draw a few objects and give a written description of what they feel and look like.

Prepare a tray of materials, which may include liquids, and ask the children to sort them according to different criteria, for example, colour or texture. They can put them into small hoops as an introduction to Venn diagrams.

Allow the children to explore materials that can change shape by drawing the different shapes that they make. Use materials such as clay, plasticine, Blu-Tack, Play-dough, pastry dough, bread dough, pipe cleaners and paper. Also, give them materials to draw that change shape, but do not retain the shape, for example, elastic bands or foam. This activity can precede an activity involving a permanent change brought about by heating, such as firing clay or baking pastry.

Ask the children to carry out an art activity with clay. It can be moulded into shape by hand or they could use pastry cutters to cut the clay into shapes. Leave the models to dry and then fire them.

Try cooking eggs in different ways, for different amounts of time. Investigate at what temperature does egg white change? Older children could drop albumen into water at different temperatures using a dropper to see at what temperature it coagulates.

Children could also make and cook pastry or a gingerbread man or woman.

Conduct a test to find out the different time it takes for different materials to melt. Place a selection of materials with the same mass into dishes. You could use wax, butter, chocolate or margarine. Place the dishes into a pan of boiling water. Which melts the fastest?

Safety : Care when handling glassware and hot and cold things. Do not allow children to taste food. See ASE publication *Be Safe!* for information on all aspects of safety in school science.

Numeracy Skills

Children should be able to :

- sort objects and materials using a Venn diagram;
- count objects in the Venn diagram;
- name and describe shapes.

Literacy Skills

Children should be able to :

- list the verbs used in the poem;
- use verbs when producing their own poem;
- spell non-standard words "ie", "ei" etc;
- understand digraphs "kn", "ph" etc.

Tank Jacket

My Dad said
the new tank
in the cupboard
needs a jacket

I thought
a jacket?
What does it need a jacket for?
It's not going out.
It hasn't got arms.
It hasn't got anything
to put in pockets

My Dad said
the new tank
in the cupboard
needs a jacket

So he went out
and brought it back,
and put in on the tank.

It didn't have sleeves.
It didn't have pockets.
The tank's not going out.

What a waste of money.

TANK JACKET

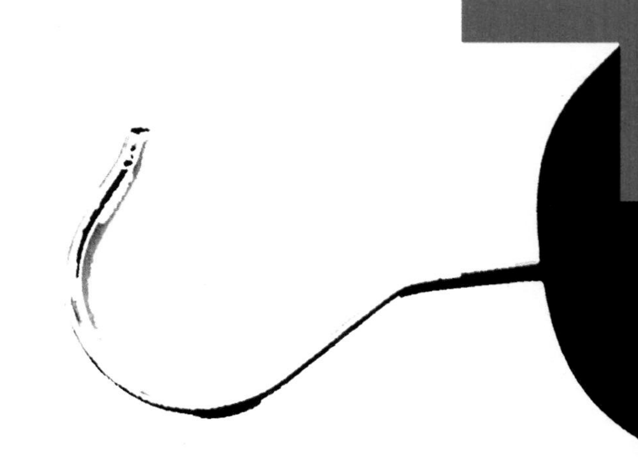

My Dad said
the new tank
in the cupboard
needs a jacket

I thought
a jacket?
What does it need
a jacket for?
It's not going out.
It hasn't got arms.
It hasn't got
anything
to put in pockets

My Dad said
the new tank
in the cupboard
needs a jacket

So he went out
and brought it back,
and put it on the tank.

It didn't have sleeves.
It didn't have pockets.
The tank's not going out.

What a waste of money.

"Tank Jacket"

Discussion Points

- What is a jacket? What does it do? Do you have one?
- What is a water tank?
- Why does a hot tank need a jacket?
- What type of materials keep you warm?
- Can we keep the cold in and the heat out? What materials will do this?
- Do you think some materials are better insulators than others?
- How do we stop cold water tanks and pipes from freezing?

Science Background

- Apart from things that produce their own heat, like fires, objects will be at the temperature of their surroundings. Things from a refrigerator or from an oven will warm up or cool down to reach room temperature. Even at room temperature, some things feel hot or cold because of their ability to conduct, that is transfer, heat energy from one object to another. Wood feels warmer than metal because it is a poor thermal conductor and doesn't conduct the heat away from your hand, but it is a good thermal insulator. Air or any material with air trapped inside such as polystyrene is also a good insulator of heat and cold. This is due to the fact that the molecules of air are not close together and so heat is transferred through the material at a slow rate. A metal is made up of particles that are packed closely together and will transfer the heat quickly through it.

Key Ideas

- Hot things cool down and cold things warm up to eventually reach the temperature of their surroundings.
- Some materials are better thermal insulators than others.
- Some materials are better thermal conductors than others.

Science Skills

Children should be able to :

- understand the concept of insulation;
- plan a simple experiment and conduct a fair test;
- take part in, or listen to, a discussion;
- work with others either in pairs or in a group.

Key Activities

Give older children beakers of hot water and/or iced water. Provide them with thermometers and timers and ask them to measure the temperature in the beakers at regular intervals.

Ask them to time the speed at which the liquid cools and ice melts. What eventually happens to both lots of water?

Retain the thermometers and repeat the experiment, this time wrapping the beakers in different materials to keep them hot or cold. Materials could include fabric, newspaper or foil.

This is a very good activity for older children to practise constructing line graphs.

It can also be used to practise datalogging using ICT sensors to take the temperature of the hot and cold water at regular intervals.

Children could also investigate different insulating containers, for example, various kinds of cool boxes, to find out which is the best. This also helps to illustrate that the same materials can keep things warm as well as cold.

Young children may need help with planning as there are many possible variables. For example, they will need to decide how often to take the temperature and the best way to record their readings. They may also need assistance using a thermometer and timer.

Very young children could investigate "which sock is best for keeping feet warm".

To do this:

1. Put some hand-hot water in small pop bottles;
2. Fasten the top securely;
3. Place each one inside a sock of a different fabric;
4. Measure the temperature at regular intervals to see which stays warmest, longest.

Safety : Recognising hazards, handling hot and cold liquids and glass thermometers.
See ASE publication *Be Safe!* for information on all aspects of safety in school science.

Numeracy Skills

Children should be able to :

- use a thermometer and stopwatch accurately;
- measure volume accurately;
- make accurate observations;
- record results accurately.

Literacy Skills

Children should be able to :

- understand terms such as verse, couplet, stanza, rhyme, rhythm;
- understand word play, for example, jacket;
- perform a poem in different ways.

Woolly Saucepan

Could I have
a woolly saucepan
a metal jumper
a glass chair
and a wooden
window-pane please?

Er-sorry I mean
a woolly chair
a glass jumper
a wooden saucepan
and a metal
window-pane
please? Could I have
a woolly saucepan
a metal jumper
a glass chair
and a wooden window-pane please?

Er-sorry I mean
Oh-blow it!
You know what I mean
don't you

Could I have
a woolly saucepan
a metal jumper
a glass chair
and a wooden
window-pane please?
Er-sorry I mean
a woolly chair
a glass jumper
a wooden saucepan
and a metal
window-pane
please?Could I have
a woolly saucepan
a metal jumper
a glass chair
and a wooden
window-pane please?
Er-sorry I mean
Oh- blow it!
You know what I mean
don't you

"Woolly Saucepan"

Discussion Points

Have a collection of objects on hand when discussing the poem.

- What are the objects in the collection made of?
- What was the saucepan in the poem made of?
- What are saucepans usually made of? Why?
- What other things are made of metal or wool or glass?
- What does metal feel like?
- What types of metal do you know?
- What does a woolly jumper feel like?
- What would a metal jumper feel like?
- What would be the problem with a metal window-pane?
- What are windows usually made of? Why?
- What is special about glass?
- What other things are made of glass? Why?

Science Background

- Children need to be aware of the diversity of uses of the same material. Making collections of different types of objects all made of the same material helps to raise this awareness. They also begin to develop an understanding that different materials are used for the same job and that some may be better than others.
- They also learn that materials may be treated in order to give them a particular property. Waterproofing is an example of this where a porous material such as cloth may be coated with a non-porous material such as plastic or wax to fill in the pores and make it waterproof. They can also begin to increase their knowledge and vocabulary of specific names such as different metals, plastics, cloth and papers and understand that they also have different properties.
- Metal is used for saucepans because it is a thermal conductor. It conducts heat well, which means it heats whatever is in the saucepan. However, because a metal saucepan handle would also get hot, saucepans have wooden or special plastic handles; these are thermal insulators.

Key Ideas

- Many materials can have a variety of uses.
- Materials have certain properties.
- Materials can be tested for these properties.
- There are special words to describe these properties.
- Some materials are good conductors of heat.

Science Skills

Children should be able to :

- follow instructions;
- plan and carry out a fair test;
- carry out experiments using different pieces of equipment.

Key Activities

Give children a collection of objects made from different materials and ask them to sort them into labelled trays. They could enter their results on Venn diagrams or on to tables. Ask them to explain their criteria for classifying the objects and challenge them to find other ways of classifying the same objects.

Give the children a selection of different types of fabrics and ask them to discover which materials are waterproof. They could do this by stretching the materials over jars and pouring water on top.

They could also find out which material dries the quickest by wetting small pieces of material and leaving them to dry.

Children could investigate thermal conductors using metal, wooden and plastic spoons using the following method.

1 Use three containers suitable for holding hot (not boiling) water.

2 Make lids for the containers, for example, using cling film. This is to prevent the steam heating the spoon handles.

3 Fill the containers with hot, not boiling water.

4 Secure the lids onto the containers.

5 Pierce a hole in the cling film lids and poke a plastic spoon, a wooden spoon and a metal spoon through the film lids into the water.

6 Use a digital thermometer or a computer sensor thermometer to measure the temperature of the handles of each spoon at the start, and at regular intervals for a set time.

Explain to the children that different materials allow heat to travel through them more or less easily. Ask them which of the materials they think will conduct heat the best. Introduce the terms "thermal conductor" and "thermal insulator".

Challenge the children to think of how thermal conductors and thermal insulators are used in kitchen equipment.

They could design objects made from silly materials such as a "chocolate saucepan" or a "jelly chair". Ask them to explain why such materials would be unsuitable.

Safety : Take care when using glassware and hot water. See ASE publication *Be Safe!* for information on all aspects of safety in school science.

Numeracy Skills

Children should be able to :

- measure capacity accurately;
- make accurate observations;
- make reasoned estimates;
- record results in chart form.

Literacy Skills

Children should be able to :

- to read and respond to humorous poetry;
- to collect words to write a poem;
- use, or construct their own, thesaurus to group words according to meaning;
- define words, for example, rough defines a property of a material.

Bottle Bank

The bottle bank gobbled up my bottle
and the bottle bank went clank.
That's bad.
Look at all the work I do
giving it hundreds of bottles to chew.
And that's not all;
at home, we've got three bins;
one for bottles,
one for paper,
and one for tins.

After all that work
I don't think a bottle bank
should say clank.
I think bottle banks
should say thanks.

The bottle bank gobbled up my bottle and the bottle bank went clank .
That's bad. Look at all the work I do giving it hundreds of bottle to chew
And that's not all, at home, we've got three bins: one for bottles,
one for paper, and one for tins. After all that work I don't think
e bottle bank should say clank. I *think* bottle banks
should say thank.

Bottle Bank

"Bottle Bank"

Discussion Points

- What is a bottle bank?
- What are they for?
- Why do you think we put different things in different bins?
- Do you think it is a good thing to recycle bottles? Why?
- Look at a bottle, a tin and some paper; describe how they resemble each other, and how they differ.
- What other things do we recycle?
- What could you recycle at school?

Science Background

- Waste management is an important local and global issue. The United Nations Conference on Environment and Development held in Rio de Janeiro in June 1992, known as the Earth Summit, developed an action plan for the 21st Century based on sustainable development. Amongst the aims was: to reduce waste, make the best use of waste, to develop waste management systems, to minimise environmental problems (Reduce, Reuse, Recycle).
- Bottle banks are just one aspect of a larger set of issues and approaches to waste management. Many children will be familiar with bottle banks but may not know about the reasons behind them and the process of recycling.
- The major ingredients of glass are sand, limestone and soda ash. These are placed in a large furnace and heated until the mixture melts to form an extremely hot liquid that can be shaped in different ways. Recycled glass is crushed and heated then shaped; this way the first parts of the process of using sand, limestone and soda ash are left out. This means that certain raw materials are not needed from the environment; they do not have to be transported and less energy is used in the process.
- The advantages of recycling bottles includes: reducing the demands for the raw materials which make glass; using recycled glass means that fewer raw materials are taken from the environment and therefore there is less damage to the environment; less energy is used in making glass from recycled material.

Key Ideas

- Materials can be grouped on the basis of simple properties.
- Materials have many properties.
- Recognise and name common materials.
- Sort and test materials for specific properties.
- Materials can be recycled.
- Recycling can be environmentally friendly.

Science Skills

Children should be able to :

- research recycling materials;
- sort and test materials for specific properties;
- carry out simple tasks;
- make accurate observations.

Key Activities

Ask children to share what they know about recycling and if they know any recycling examples in school, at home and in the locality.

Bring in clean waste from the teacher's home; this is waste which has been cleaned, for example, tins and bottles washed out, tea bags, vegetable peelings, newspapers and card boxes such as cereal packets.

Children do not have to handle this waste but they could decide how each one could be recycled. Discuss the various ways with children, such as taking glass, newspaper and tins to recycling points, placing tea bags and vegetable peelings on a compost heap and cereal boxes brought to school for technology, science and art activities.

Samples of the clean waste can also be buried in containers of soil and left outside or buried in a plot of soil in an area of the school grounds that is safe and not subject to fouling by dogs. At regular intervals dig the items up, make a note of changes, then re-bury them.

Challenge children to create a poster or a leaflet giving information about materials and recycling to other children or to parents. Display the children's work around the school.

Children could be introduced to the recycling code which asks people to:

rinse bottles, jars etc;

remove tops of bottles;

place glass in the correct bin;

take other things to recycle such as newspaper, clothes and tins so that only one journey is made and less energy is used.

Safety : See ASE publication *Be Safe!* for information on all aspects of safety in school science.

Numeracy Skills

Children should be able to :

- use and make Venn diagrams;
- sort items into sets;
- count objects in a set.

Literacy Skills

Children should be able to :

- understand the use of rhyming words;
- read a variety of poems on a similar theme, for example, recycling;
- compose their own poem using a repetitive pattern.

By Pass

All the people in the town said:
we want a by-pass.
Send the road
round the town
not right through the middle

Our children are being
knocked by the traffic:
people are getting asthma,
the buildings are falling down.

So they built the by-pass.

It went through the middle of a wood.
They knocked down trees,
the birds and animals have gone
and the cars and lorries
go whoosh whoosh whoosh

Though sometimes
they go
whoosh whoosh crash

Related poems:

Dirty T-shirt
How Humans Out Died

By Pass

All the people in the town said:
we want a by-pass.
Send the road
round the town
not right through the middle

Our children are being
knocked by the traffic:
people are getting asthma,
the buildings are falling down.

So they built the by-pass.

It went through the middle of a wood.
They knocked down trees,
the birds and animals have gone
and the cars and lorries
go whoosh whoosh whoosh

Though sometimes
they go
whoosh whoosh crash.

"By Pass"

Discussion Points

- What is a bypass?
- Do you know of a bypass? Where is it? Why was it built?
- What changes took place because the bypass was built?
- Who decides to build a bypass?
- What are the good and bad things about a bypass?
- Where do you think the birds and animals go when a bypass is built?

Key Ideas

- Pros and cons of environmental improvements – damage to ecosystems versus other issues such as public health, the need to reduce traffic accidents in towns etc.
- Air pollution is caused by human activity and causes damage to both the environment and its inhabitants.

Science Background

- When deciding whether a bypass should be built there are usually two sides to the argument. People in many areas of the country are in favour of a bypass, usually because of traffic problems such as the volume of cars and large lorries going through a small village or town, and noise and other pollution issues relating to safety of residents.
- In terms of pollution, road traffic contributes the following to the atmosphere: nitrogen oxides; benzene – which plays an important role in smog; carbon monoxide – which, as a gas, deprives the body of oxygen and can cause headaches; sulphur dioxide – which can cause tightening of the airways and aggravate asthmatics and people suffering from bronchitis; dust particles – which are emitted as part of black smoke from vehicles and are associated with heart and lung diseases.
- Linked to pollution is public health; there are approximately three million asthmatics around the country and it is known that pollutants from road traffic can aggravate and trigger asthma attacks.
- However on the other side of the argument there may be people who consider the environmental impact too devastating to warrant this type of construction. Many people are concerned about losing ecosystems, areas of special scientific interest (SSIs), plants and animals, as well as losing green belt land and land which is considered beautiful. Traffic can have a direct and indirect effect on wildlife. Rabbits, foxes, pheasants and hedgehogs (as many as 100,000) and amphibians, as well as badgers (approximately 47,000) are killed by vehicles on the roads. When a bypass is built, land which housed a range of habitats for plants and animals disappears.

Science Skills

Children should be able to :

- choose and use appropriate apparatus, for example, filter papers, funnels and correct sized containers;
- decide what they need to know and plan an experiment to find the answers;
- understand the purpose of a control experiment;
- collect and evaluate evidence from various sources;
- work with others.

Key Activities

Prepare laminated cards, plastic sheeting or glass slides covered with double sided sticky tape or Vaseline and hang them in areas of the school grounds or the near vicinity. Children could predict which will show high and low levels of pollution, for example: a car park, the road or the garden. Leave for one or two days and look at the deposits with a magnifying glass. What do their results tell them? Children could take one home and compare the results in school. The following question could be set as an investigation. If a wet t-shirt was left to dry in or around your school, where would it get most dirty?

A teacher demonstration can be set up to show the impact of sulphur dioxide on plants.

1 Crush a Camden Tablet (obtainable from chemists)
2 Add two tablespoons of lemon juice in a glass jar.
3 Take two pots of cress.
4 Put one pot of cress and the jar in a plastic bag and fasten so the bag is sealed.
5 Put the second pot of cress in a sealed jar.
6 Observe the effects on the cress.

The gas given off by the Camden Tablet is sulphur dioxide and turns the cress yellow. Make a daily diary for a week.

Sample evergreen leaves, such as holly of the same age (count rings on the stem) from different locations. Wipe the leaves with absorbent white paper and compare the deposits. What explanation can be given for the findings? Ensure the place the leaf is taken from is recorded.

Ask the children to collect rainwater in a bottle with a filter paper and funnel placed in the top. Do the same by pouring the same amount of tap water through a filter. Compare any deposits with a magnifying glass.

They could also grow cress with distilled water and also with water and vinegar mixed in different proportions to see the effect on the cress. This simulates acid rain.

Either as a demonstration or as a supervised activity, burn small samples of safe pre-tested material to observe and compare the amount of smoke given off (wear goggles). Relate this to the impact on the environment of much larger scale burning.

Observe the deposits left on an inverted spoon held over a burning candle flame; or use filter paper or cotton wool placed in the far end of a jar over a burning candle to collect deposits and observe them.

Safety : Wear safety goggles during burning experiments.
See ASE publication *Be Safe!* for information on all aspects of safety in school science.

Numeracy Skills

Children should be able to :

- measure capacity accurately;
- record data in a table;
- understand the concept of scale;
- relate results to larger or smaller scale;
- represent data as a graph.

Literacy Skills

Children should be able to :

- use persuasive arguments;
- write a letter to complain about a bypass;
- use computer technology to produce the letter in an appropriate layout or format.

Giants

As we drew near
we could hear
the thudding
of their hearts
beating.

The moment
we appeared
they stood to attention;
row upon row
across the hillside,
their great white arms
revolving.

It was obvious.
They were communicating
with each other
before making their next move.

It will be
fascinating
to see how they
manage to walk
down the hill
on one leg.

Giants

As we drew near
we could hear
the thudding
of their hearts
beating.

The moment
we appeared
they stood to attention:
row upon row
across the hillside,
their great white arms
revolving.

It was obvious.
They were communicating
with each other
before making their next move.

It will be
fascinating
to see how they
manage to walk
down the hill
on one leg.

"Giants"

Discussion Points

- Why is the poem called "Giants"? What do you think "they" look like?
- Why do you think the poem says "The moment we appeared they stood to attention"? Did they move? How many do you think there were "across the hillside"?
- What were "their great white arms" and why were they revolving?
- Do you think they were communicating with each other? Why does the poet think it was obvious?
- How were they making a sound like "the thudding of their hearts beating"?
- Do you think they are likely to walk down the hill on one leg? Why "one leg"?
- What do you think they were really on the hillside for? Was the hillside important? Why?

Science Background

- The force of the wind can be used to do work for us. This happens when the wind acts upon, or blows against, a mechanism, such as a windmill or a turbine, causing it to move – usually to turn. This turning force can then be directed, usually through other moving parts like cog-wheels and shafts, to the part of the mechanism that produces the useful work like grinding corn, lifting water or generating electricity.
- Energy, which can be thought of as the capacity for doing work, is transferred from the wind, which is a moving mass of air, through the various moving parts of the mechanism to the final working part where it appears as useful work to us. The amount of energy available as useful work is less than the amount transferred from the wind because each moving part transfers some of the energy to heat through friction. However, overall the amount of energy is constant – there is no loss or gain, but not all of the energy from the wind is turned into useful energy.
- Although wind power is a renewable source of energy, windmills are not all good news. Local residents sometimes object to them on the ground of noise pollution and visual intrusion.

Key Ideas

- The force (energy) of the wind can be used by us.
- There are alternative sources of energy.
- Some sources are renewable.
- Scientific and technological developments have both positive and negative effects.

Science Skills

Children should be able to :

- test a hypothesis;
- make notes from various sources, including secondary sources;
- organise information and findings for report and display;
- write explanations and explain their findings to others;
- consult books and databases.

Key Activities

Note: Careful supervision may be needed for this. Take children outside on a windy day to try to catch the wind using different surfaces. They could use their jackets, sheets of card and hardboard, toy windmills and model aeroplane propellors. These could have been discussed in the classroom beforehand. Firstly, get them to stand still in the wind and ask them what they feel. Then tell them to hold their various surfaces up and ask them what they feel. What happens to the force of the wind if they turn to face in different directions? Are there any directions where it feels stronger? How does this relate to the direction in which the wind is blowing?

Back in the classroom, children could think about which surfaces would be best for using the force (energy) of the wind. They could consider what they could do with this captured force.

The oldest children could investigate different windmill designs to see which is most effective. This could be extended by researching into different windmill designs using books, data-bases, etc.

If possible, visit a/some working windmills and/or a wind turbine to see how the energy of the wind can be used to do something useful for us. The children should ask questions about how each windmill was designed to make the best use of the force (energy) of the wind to suit that windmill's particular purpose.

They could find out how the wind is caught. What happens when the wind changes direction? How is the wind's energy used to turn grindstones, generate electricity, etc?

Again this could be extended, or even done, by researching books, local guide-books, environmental publications, data-bases, the Internet, etc, but be careful to specify the relevant questions. The issue of wind as an alternative source of energy might be included. This would be a very good opportunity to consider the limitations of this sort of research, and the value to be placed on the evidence collected.

The use of different types of windmill in different countries, particularly developing countries, could be explored. What are the advantages and disadvantages of wind power in these situations?

Challenge the children to design and make a model windmill or other wind-powered device.

Safety : No specific hazards relating to the science activities or resources, but the usual safety precautions for children carrying out activities outdoors and going on visits.
See ASE publication *Be Safe!* for information on all aspects of safety in school science.

Numeracy Skills

Children should be able to :

- collect data, for example, data about the weather;
- use tables, charts and graphs;
- calculate mean, mode and median, for example: for temperature or wind speed.

Literacy Skills

Children should be able to :

- understand figurative language in poetry;
- continue the poem using a similar form;
- write a poem on a similar theme;
- research alternative energy using non-fiction books.

I Would Die Without

One man said
I would die
without TV

and the other said
I would die
without food.

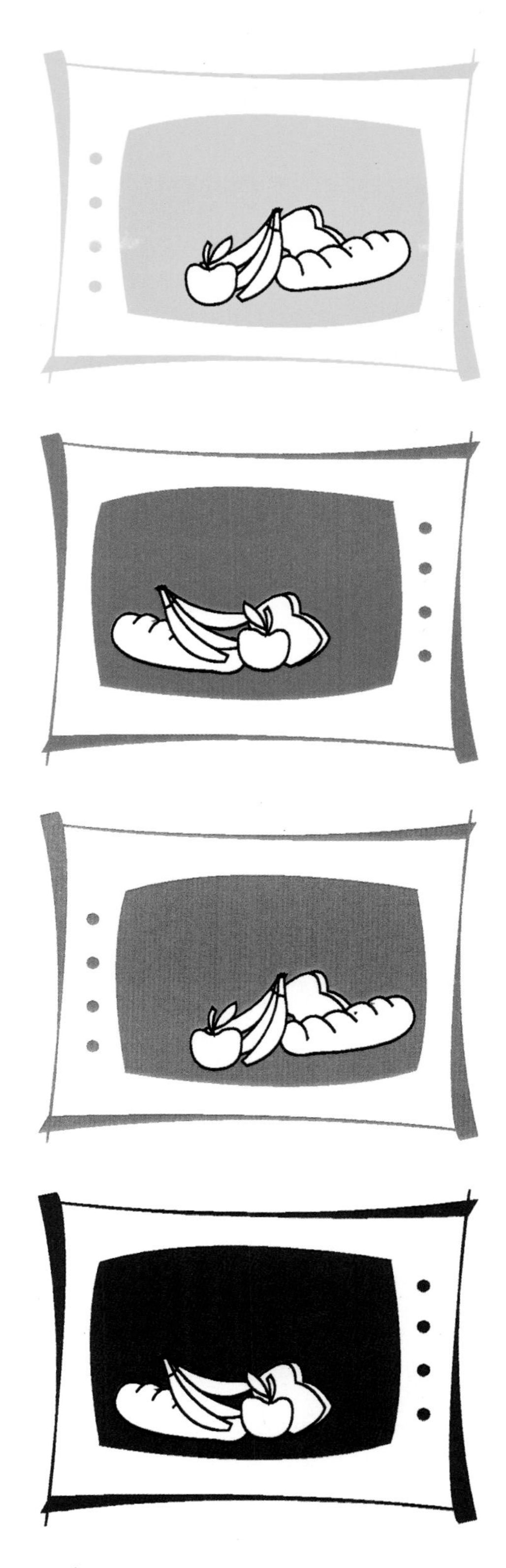

I would die without...

One man said
I would die
without TV

and the other said
I would die
without food.

"I Would Die Without"

Discussion Points

- Would you die without TV?
- What did the man who said this really mean?
- Would you die without food?
- Do you think that everyone has enough food to eat in the world? How do you know?
- What kind of things can happen to people if they do not get enough food?
- What do we need to stay alive? Can we live on food alone?
- Do you think we eat too much food?
- What kinds of food could you give up?
- What kinds of food should you include in a balanced diet?

Science Background

- A correctly balanced diet is one that allows animals – including humans – to stay healthy.
- A balanced diet must also contain essential vitamins, minerals and water.
- The three main food groups are carbohydrates, proteins and fats.
- The body uses carbohydrates and fats as an energy source.
- Fats are very high in energy provision.
- Starch provides long-term energy as the digestive process takes a while to convert it into glucose and enter the blood stream.
- Sugar is absorbed into the body without going through the digestive system and so provides immediate energy.
- Excess starch and sugar is stored as fat.
- Proteins, when digested, produce amino acids.
- Amino acids are used for growth.
- Water is vital to maintain the water in the tissues. Even moderate water loss can be very dangerous.

Key Ideas

- Some foods are better for us than others. Fruit and vegetables help you to keep healthy.
- Water is needed to keep us alive.
- Different foods do different jobs in the maintenance of the body.
- For food to be useful it must be broken down by the process of digestion.

Science Skills

Children should be able to :

- use secondary sources of information;
- understand there are different sorts of food and they are used in the body in different ways;
- research different food groups.

Key Activities

Let children sort foodstuffs into foods they like and dislike. Do they have favourites in common? Why do they like them?

Let children sort foodstuffs into foods that they think they should eat a lot of and those they should eat little of. Do they recognise fruit and vegetables and do they realise why they need to eat them? Allow children to taste a variety of cut raw fruits and vegetables including less familiar ones.

Discuss food values with the children. Ask them to place food on a traffic light system, for example, RED foods (STOP – eat very little), amber (OK), green (GO – eat plenty).

Do a survey of drinks the children like. Can they group them into juices, fizzy drinks, hot drinks, squash? Show the children the sugar contained in them. Do they know that water is the main content of all drinks? Look at the drinks' labels to see what they contain.

Discuss with the children who chooses the food they eat? What influences affect their food choice, for example: beliefs, advertising, free gifts, preparation time, health, cost, looks, taste?

Ask the children to group foods according to their own criteria. Discuss and assess their understanding of the groups. Using secondary sources regroup them according to food types.

Use food packages to explore the energy in foodstuffs. Measure out sugar quantities and fat quantities, using lard for visual understanding of amounts involved.

Ask children to draw an annotated diagram of how they think food gets from the mouth to the muscles. Where would a drink of water go? Children will often annotate a water pipe and food pipe. Use the annotated drawings to identify which organs the children are familiar with and any common misconceptions. Use secondary sources to explain the journey of food through the digestive system: include teeth, tongue, saliva, gullet, stomach, intestine.

Safety : Whenever children handle food, even if it is not eaten, they should be encouraged to follow normal hygiene procedures. Make it clear to the children that they should only eat food in the classroom with permission.
See ASE publication *Be Safe!* for information on all aspects of safety in school science.

Numeracy Skills

Children should be able to :

- use numerical information on food packets;
- carry out a survey on food preferences;
- produce graphs using data from the survey.

Literacy Skills

Children should be able to :

- analyse how messages are conveyed through poetry;
- use non-fiction to research information about diet;
- represent information in different ways, for example, as an advertisement, or a play, or an information leaflet.

Save It

When we're in school
they say to us
we should do all we can
to save energy;
use less electricity, petrol and gas;
use less water.

I don't know why
they keep telling us!
They ought to tell
all those people
in the ads for
cars, cookers, petrol, washing machines
and stuff...

Save it

When we're in school
they say to us
we should do all we can
to save energy:
use less electricity, petrol and gas;
use less water.

I don't know why
they keep telling us!
They ought to tell
all those people
in the ads for
cars, cookers, petrol, washing machines
and stuff...

"Save It"

Discussion Points

- What different kinds of energy are mentioned in the poem?
- Can you think of any objects that use these different kinds of energy? Can you see any in the classroom?
- How is electricity used in your school?
- How could you help save electricity at your school?
- Who does the child in the poem think is to blame for using too much electricity?
- What could these people do about the problem?
- Can you think of different ways to save electricity at home?
- When the washing machine is on, where is the energy transferred to?
- What do you think an advertisement should say to encourage people to save energy?

Science Background

- Energy cannot be used up, but an energy resource can.
- Energy can be dissipated, or spread out, so that it is not useful or useable. A simple example is a gas cooker where the energy resource, gas, is burned and used for cooking. The useful energy is transferred to the pan, heating it and its contents. But some of the energy in the burning gas is transferred to the top of the cooker, and to the surrounding air. This energy is not useful.
- The main non-renewable energy resources are effectively fuels, and are burned whole, or in parts. Their energy is associated with the system fuel plus oxygen because the energy is not available unless the fuel burns. Once the resource (fuel) has been burned, it cannot be renewed.
- The extraction of an energy resource can cause direct damage to the physical environment, and the process of obtaining useful energy can cause indirect damage by polluting the atmosphere and affecting the growth of living things.
- However, there are different opinions about the amount and extent of this potential damage.

Key Ideas

- Non-renewable energy resources can be used up, which may damage the environment.
- More care can be taken when using energy sources, for example, in the home.
- There are positive and negative effects of scientific and technological development on the environment.

Science Skills

Children should be able to :

- listen to different opinions from others;
- offer their own reasoned opinions;
- take part in discussion;
- report findings to others;
- work in groups.

Key Activities

Note: There are no suitable practical activities but many opportunities for children to research different aspects of energy use, individually and in groups. Be aware that a wide range of opinions and viewpoints are an essential feature of this general area. It is important that this range is represented in the children's work.

Younger children could find out what the main energy resources are: oil, natural gas and coal. They could also find out where they come from, how they are extracted and what happens to them to produce useful energy.

Older children could begin to explore the environmental implications of extracting these resources and consider if there are any alternative or renewable resources.

They might also look at the waste products from using oil, natural gas and coal and find out what effects they are considered to have on the environment.

The oldest children could extend these aspects and look at different opinions put forward by interested groups such as the energy suppliers, the consumers, the environmentalists and the Governments.

These areas of research will involve children using books, newspapers, information supplied by interested parties, data-bases, the Internet, TV, etc.

Children could research and list the use of electricity in school. They could be given access to the school electricity bill and plan a "save it" campaign and evaluate whether the campaign was effective. For details of the "Young Energy Savers" scheme, contact Groundwork on 0121 236 8565

Local industry and businesses could be approached for information on their energy saving policies. These could be compared with the children's own ideas.

The whole area provides excellent opportunities for inviting speakers to school or for making visits.

Safety : The usual care must be taken when children are going on school trips.
See ASE publication *Be Safe!* for information on all aspects of safety in school science.

Numeracy Skills

Children should be able to :

- collect information from secondary sources;
- assemble data in different formats;
- display data in different ways.

Literacy Skills

Children should be able to :

- read and interpret a poem with multi-layered meanings;
- produce a poster about saving energy;
- write a letter to a local environmental group asking for information;
- write an energy information leaflet aimed at different audiences, for example, younger children or grandparents.

star*

◆ related poems

Centrally Heated Knickers

I know a man who collects stickers,
like: 'I am wearing centrally heated knickers.'

If it's true, and he's not a liar
I might see him some time with his pants on fire.

Dirty T-shirt

My clean T-shirt was dirty.

OK. This is how it goes:
first it was dirty.
Then I washed it.
Now it's clean.
So I put it on the washing-line,
and it dries
and now it's dirty.

That's my dirty, clean T-shirt.
Tiny specks of dirt on it.
But why?
Is someone throwing the dirt?
Is the wind blowing the dirt

So I think:
I'll spy on my T-shirt.

The next day

I washed it again,
I put it on the washing line
and I watched...

...no one threw anything at it
The wind didn't blow. ..

then it rained
Ah-hah!

I walked over to the T-shirt.

The Doorstep

The first cold bite
of the winter wind
sinks in
just as you tread
on to the front doorstep
where the cat
has snuggled up
to the front door
to make herself warm

Engine Oil

My dad said to his friend
'I can't find my funnel.
Can I borrow yours?'
'Funnel?
What d'you need a funnel for?'
'For the oil.
To put the oil in the car.'
'You don't need a funnel for that.
I'll show you.'

He grabbed a plastic bottle
and cut the bottom off it.

'There's your funnel.'

My dad
put the open top of the bottle
down into the hole
for the oil
and the thick yellow syrup
globbed and glooped
through the bottle
into the engine
like it knew the way there
from long ago.

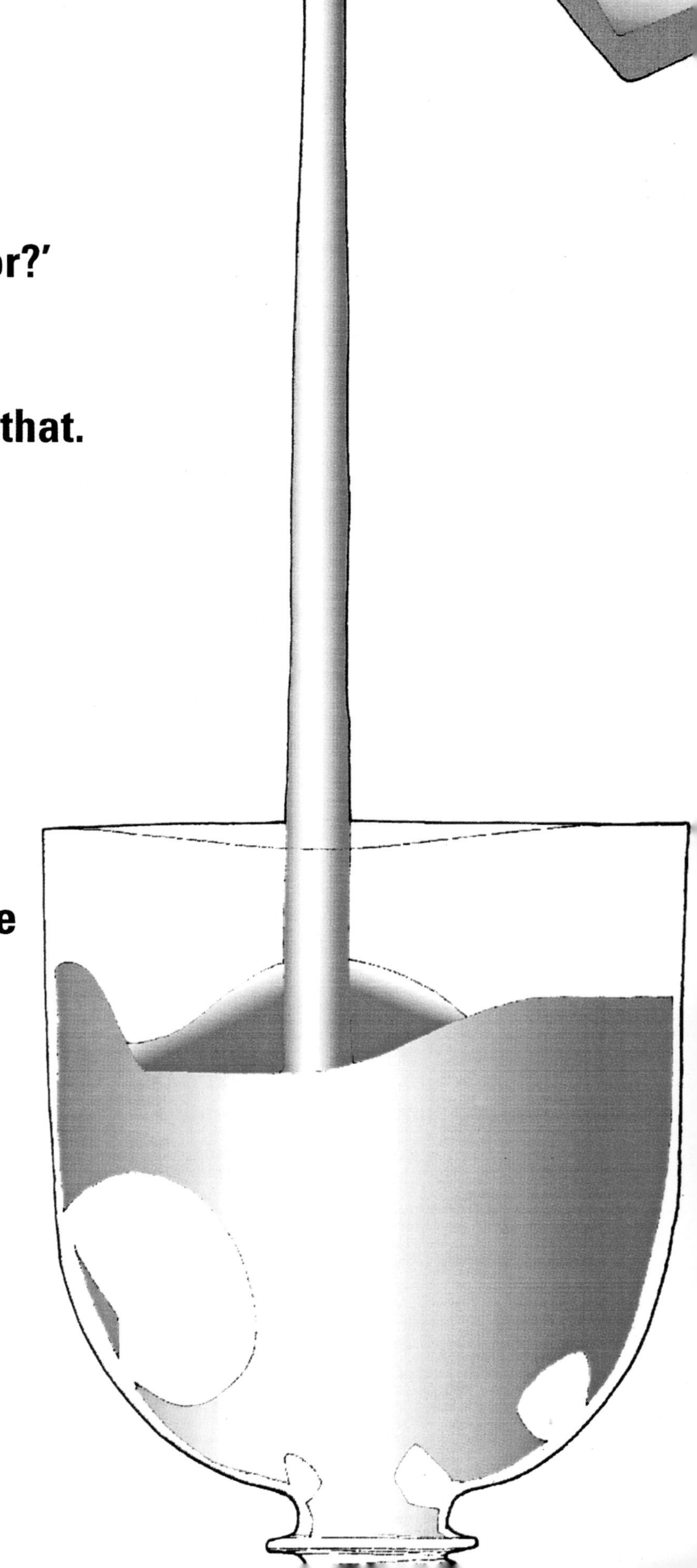

Growing Apples

And the King said,
'How do I turn this apple
into thousands of apples?'

The wise men scratched their heads,
muttered amongst themselves
and consulted their great books.

One stepped forward.
'Perhaps this is some kind of joke,
your majesty,
but could one say
that one could make
a thousand apples
by chopping one apple into a thousand pieces?'

'Balderdash!' said the King
'I said thousands of apples
not some nonsensical business
of hacking an apple to bits.'

Another wise man stepped forward.
'I have heard that beyond the horizon
there lives a man
who sings to the objects in his house
it is said of him
that he can cause things to multiply.
Maybe-'

'Poppycock!' roared the King,
'I wasn't looking for some holy-moly jigger[illegible]

And on it went.

None of the wise men
were wise enough to solve the problem.

A serving-girl
who was pouring the wine
caught the drift of what was going on.

'I know how to turn your apple
into thousands of apples,' she said.

How the wise men laughed!
'The cheek!'
'A little whipper-snapper like her!'
'As if she'd know!'

'Come then,' said the King
'Speak, girl!'

'I would bury your apple.'
Said the girl.

There was silence.

The wise men looked at each other
and sniggered.
'Bury it? Bury it?
What good would that do?'

But the King didn't wait
'You're right, young lady.
Completely and utterly right.'

Hot Cat

The butter melted
the cheese went mouldy
it all got so hot
the cat moulted.

Then it got hotter
the cheese melted
and the cat went mouldy.

Then it got hotter
and the cat melted.

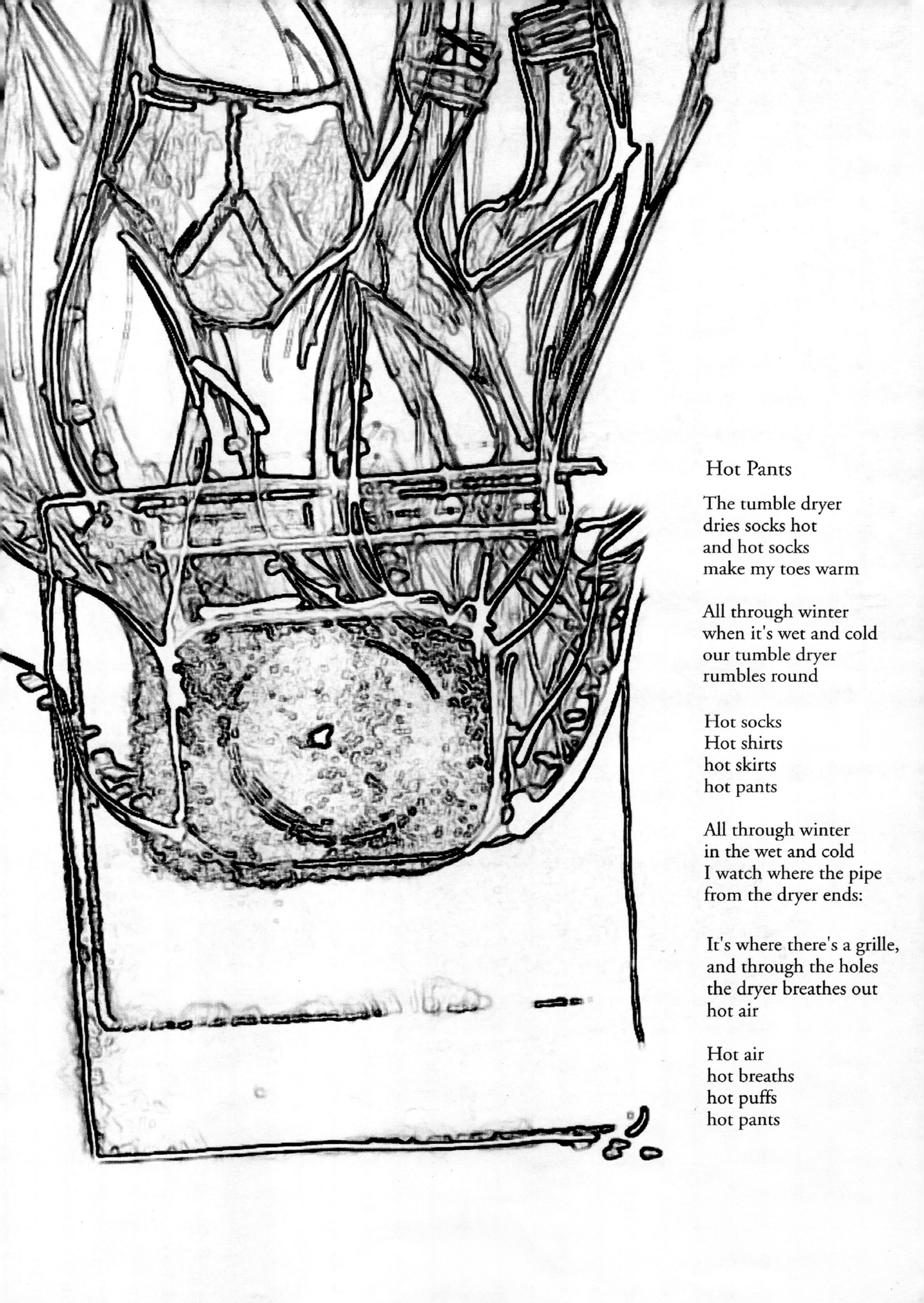

Hot Pants

The tumble dryer
dries socks hot
and hot socks
make my toes warm

All through winter
when it's wet and cold
our tumble dryer
rumbles round

Hot socks
Hot shirts
hot skirts
hot pants

All through winter
in the wet and cold
I watch where the pipe
from the dryer ends:

It's where there's a grille,
and through the holes
the dryer breathes out
hot air

Hot air
hot breaths
hot puffs
hot pants

How Humans Out-died

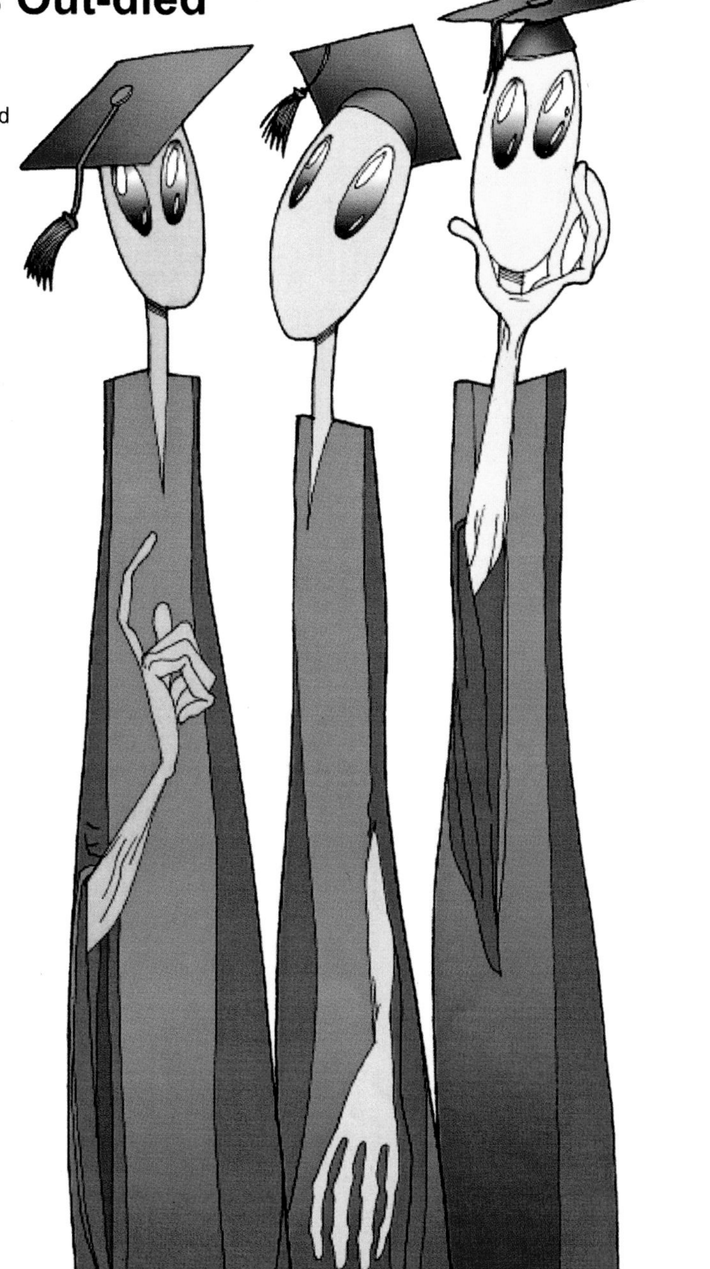

When the Martians landed
in the world they'd been handed
there were no humans left.

They said, 'Do know we how
humans did how out-die?
Mystery is it? Or why?'

One clever young Martian
stepped forward:

'I how know
that humans out-died:
much car
much carbon
much carbon-dioxide.
Exhausted by exhaust;
they car-bon diox-died.'

At that,
another clever young Martian
stepped forward and said:

'Oh no!
Not so!
No oh!

No more rain it was.
That was a pain it was.
That was a strain it was.

So,
it was not so
that humans died out.
What it was,
was that
humans dried out.'

But then a third clever young Martian
stepped forward and said:

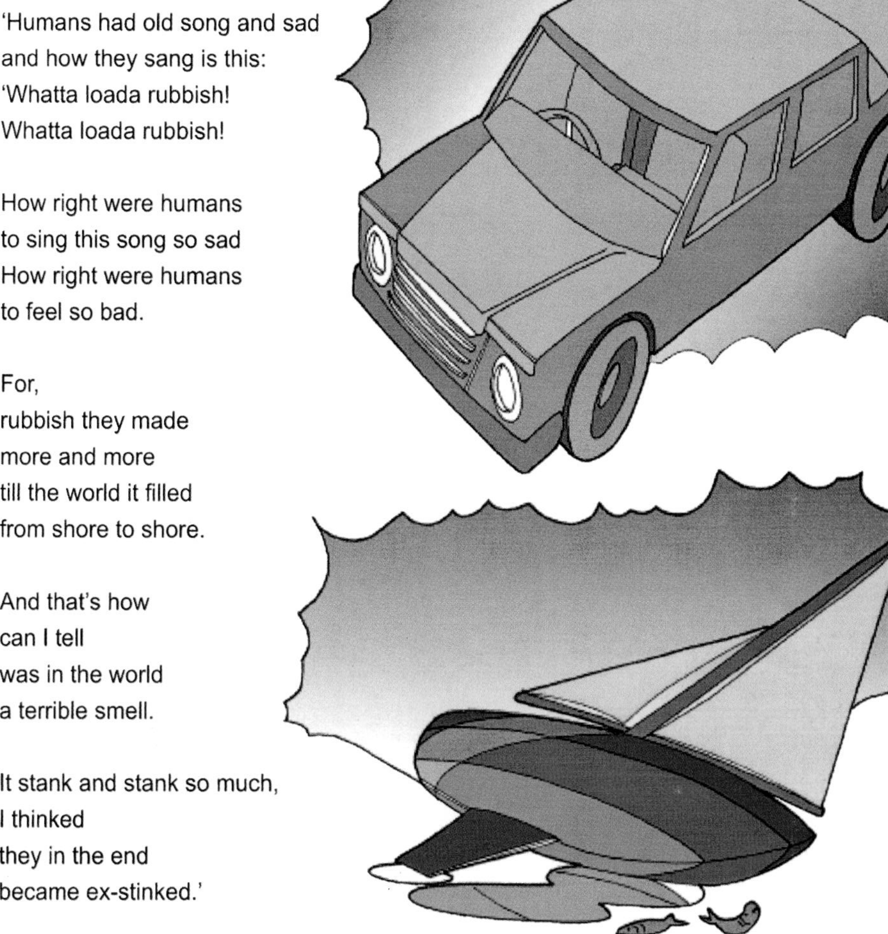

'Humans had old song and sad
and how they sang is this:
'Whatta loada rubbish!
Whatta loada rubbish!

How right were humans
to sing this song so sad
How right were humans
to feel so bad.

For,
rubbish they made
more and more
till the world it filled
from shore to shore.

And that's how
can I tell
was in the world
a terrible smell.

It stank and stank so much,
I thinked
they in the end
became ex-stinked.'

And the Martians all said:

'How right that sounds,
now know we the history
how humans out-died.
Now solved is the mystery.'

Lubricate the joints
and the railtrack points
the gears
and the shears
and the clocks
and the locks;
the drills
and the mills
and the trimmers
and the strimmers
and the rotors
and the motors
keep them whirring
keep them purring
keep them smooth
on the move
in the groooooooooooooove
oh yeah!

OIL

Some thoughts about eggs...

1. Is a hard-boiled egg one that was hard to boil?
2. How fast do runny eggs run?
3. It can't be very fast because my dad beat one.
4. If the white bit's called the 'white', why isn't the yellow bit called the 'yellow?'
5. If a piglet is a little pig, is an omelette a little om?
6. What is an om?
7. You put mushrooms in mushroom omelettes, you put cheese in cheese omelettes. I'm worried about what they put in Spanish omelettes!
8. Why don't eggs melt when you heat them up?
9. Who lays Easter eggs?
10. If we had an egg we could have egg on toast - if we had some toast.

THIRSTY LAND
In the plane over the desert
I see the badlands beneath us
wrinkled like old skin.
Ancient river valleys
are now dry brown snakes.
The sun that glares at this
has dried up
every last drop.
Even the rocks begin to say:
'Water, please, water.'

Voolly Hats

n winter

vhen we go for walks
ve take hot drinks
n flasks
nd we bury them deep
n our bags
vrapped up in woolly hats

n summer

vhen we go for walks
ve take cold drinks
n flasks
nd we bury them deep
n our bags
vrapped up in woolly hats

index 1

index 2